Publishing Opportunities for Energy Research

Publishing Opportunities for Energy Research

A Descriptive Guide to Selective Serials in the Social and Technical Sciences

Compiled by

Roberta A. Scull

Center for Energy Studies, Louisiana State University

Bibliographies and Indexes in Science and Technology, Number 1

Greenwood Press
New York • Westport, Connecticut • London

Library of Congress Cataloging-in-Publication Data

Scull, Roberta A.
Publishing opportunities for energy research.

(Bibliographies and indexes in science and technology, ISSN 0888-7551 ; no. 1)
Includes index.
1. Power resources—Research—Periodicals—Bibliography. 2. Power resources—Research—Authorship. I. Title. II. Series.
Z5853.P83S38 1986 [TJ1632] 016.33379 86-14975
ISBN 0-313-25160-6 (lib. bdg. : alk. paper)

Library of Congress Catalog Card Number: 86-14975
ISBN: 0-313-25160-6
ISSN: 0888-7551

First published in 1986

Greenwood Press, Inc.
88 Post Road West, Westport, Connecticut 06881

10 9 8 7 6 5 4 3 2 1

CONTENTS

ACKNOWLEDGMENT

There are many people to whom I owe a great deal of appreciation in seeing this project through to final publication. Primary among these is Deborah August, who, with me, has seen this work through from start to finish. She tenaciously mailed, sorted, and filed questionnaires and single-handedly entered every entry, including edited corrections, into our university computer. Without her this project might never have seen completion, and it is to her I dedicate this work.

Others who were especially helpful are two graduate students, Ivy Prewitt, who assisted me during the initial stages, and Judy Harlow, who doggedly tracked down and verified final information.

And, of course, I am especially thankful for the cooperation of the many editors who responded to the questionnaire, supplied us with journal samples, or talked with me over the phone. Without their time and interest, this publication would not have materialized.

On the production side of this publication, the Division of Research Services (DRS) at Louisiana State University, and particularly staff programmer Randy C. Hebert, made significant contributions. By using techniques developed specifically for this project, the DRS automated production of the Index and worked with LSU Graphic Services in automating the phototypesetting of the GUIDE itself, thereby saving long hours of arduous cross-referencing, typesetting, proofing, and editing.

As is often the case, a publication of this nature required the support, if not the actual assistance, of many, especially those who work with me daily at the Center for Energy Studies. To all of my co-workers I extend a special appreciation for your support and encouragement in this endeavor.

PREFACE

Purpose

An efficient and orderly cycle from research to publication advances scientific knowledge on all fronts. Such a cycle exists in traditional academic disciplines—biology, chemistry, engineering, history, physics, political science,etc.—but it does not exist in the multi-disciplinary field of energy research.

Energy researchers are often unsure of where to publish their research if it is not suitable for the disciplinary journals with which they are familiar. As a result they must search for other publication channels. Since journals tend to be indexed by discipline, an author's search for the "right" publication channel, subsequent submission, and eventual publication is a problematic process with negative and far-reaching side effects. Articles tend to appear less timely than they should; and hence, the accumulating knowledge that science aims to build is unnecessarily delayed.

PUBLISHING OPPORTUNITIES FOR ENERGY RESEARCH identifies alternative titles for consideration and facilitates the energy researcher's quest for a publication channel. Hence, it contributes to more timely publications of energy research results and speeds the entire research building process.

Scope and Selection Methodology

The "Master List" includes 524 foreign and domestic English language multi-disciplinary publications which contain, among other things, energy-related research results in aeronautics, agriculture, architecture, biology, chemistry, economics, engineering, environmental sciences, geosciences, physics, and the social sciences. The primary criteria for inclusion in the "Master List" are: (1) the publication must accept unsolicitied manuscripts in English and (2) at least partially be published in English.

Prospective titles were ascertained by examining ULRICH'S INTERNATIONAL PERIODICALS DIRECTORY; ENERGY INFORMATION ABSTRACTS (EIC/Intelligence, Inc., N Y, N Y) and ENERGY RESEARCH GUIDE: JOURNALS, INDEXES, AND ABSTRACTS by John Viola, Newell B. Mack, and Thomas Stauffer (Ballinger Publishing Co., Cambridge, MA). Once a core list was developed, subject experts and researchers were consulted regarding titles to add or delete. Titles which were considered extremely well-known and single disciplinary were generally excluded. Conversely, less known titles with the possibility of multi-disciplinary coverage were added. These efforts resulted in a final list of approximately 750 publications which were then surveyed with initial and follow-up questionnaires and, in some cases, by personal phone calls. Frequently a current edition of the publication itself was analyzed for additional information. Titles and information which were verified by respondents comprise the bulk of the "Master List" section.

Prior to publication all titles were verified and additional information obtained from THE 1985 IMS/AYER DIRECTORY OF PUBLICATIONS; THE STANDARD PERIODICAL DIRECTORY, NINTH EDITION, 1985-86; ULRICH'S INTERNATIONAL PERIODICALS DIRECTORY, 1985; or the OCLC information network. Every effort has been made to provide the most current information.

Many titles were deleted from the "Master List" after careful consideration for a number of reasons. Among the most common of these were: (1) basic information, such as whether or not unsolicited manuscripts were accepted, could not be determined; (2) the publication did not accept unsolicited articles; (3) articles were solely translations from non-English language publications; (4) the publication seemed to be of no use to researchers as a way of publishing or announcing their research efforts; (5) the publication no longer existed. These deleted titles comprise Appendix I, "Periodicals Surveyed But Excluded."

Arrangement and Accessibility

The "Master List" is arranged alphabetically by title and includes the following elements: journal address, circulation, frequency of publication, the editor's name and address, telephone number of the editor or publisher, the ISSN which is the International Standard Serial Number, the number of manuscript copies to be submitted for review, the review period (Peer review is indicated by PR.), publication lag time , style sheet requirements, acceptance of revised theses or dissertations, types of articles accepted, the major content areas of the journal, and miscellaneous notes of importance. Each category is explained in "Questionnaire Sample and Definitions."

The "Master List" is accessed through the "Subject Index" which is based on standardized U. S. Department of Energy subject categories as listed in the U. S. Department of Energy, Office of Scientific and Technical Information, ENERGY DATA BASE: SUBJECT CATEGORIES AND SCOPE, March 1985. References from the "Subject Index" are to numbered entries in the "Master List." A * following a number indicates that that particular title incorporates a policy of peer review. Since detailed subject categories have been assigned to each title, access is at the maximum. It is this outstanding index which makes the PUBLISHING OPPORTUNITIES FOR ENERGY RESEARCH such a useful and usable tool.

Following the "Master List" and the list of "Periodicals Surveyed But Excluded," (Appendix I), is a section of "Style Manuals for Energy Authors," (Appendix II). This provides complete bibliographic information on all style manuals which are referenced in the "Master List."

Appendix III is a reprint of the U. S. Department of Energy Office of Scientific and Technical Information publication ENERGY DATA BASE SUBJECT CATEGORIES AND SCOPE, March 1985. This information is provided in detail in order to provide the researcher with a good overview and description of the areas covered by the "Subject Index."

QUESTIONNAIRE SAMPLE AND DEFINITIONS

The following questionnaire was sent to editors of titles which were being considered. The accompanying letter stated that a response was requested only if the journal accepted unsolicited English language articles for publication. The initial response rate was fifty-eight percent. A follow-up questionnaire was mailed to non-respondents. This resulted in a seventy-two percent overall questionnaire response.

Information from returned questionnaires was computerized and returned to the editor for verification. Accompanying the printout was the "Explanation of Each Category." Our letter requested signed corrections or approval; but it also stated that a non-response would indicate approval.

Sample Questionnaire

QUESTIONNAIRE FOR ENERGY RESOURCE GUIDE

Journal title

____

____

Editor's Name and Telephone No.

____

Manuscript Address

____

____

____

____

Type of articles accepted (research, project, review, etc.)

____

____

____

____

____

Major content areas (topics of interest to the journal)

____

____

____

____

____

No. of manuscript copies required for review ____________________

Review period ________________________________

Peer Review yes ________ no ________

If so, which style sheet to follow? ________________________

____

Are revised theses and dissertations accepted? yes ________ no ________

Reprint policy yes ________ no ________

Circulation __________________________________

Frequency __________________________________

Miscellaneous: ________________________________

____

____

Explanation of Each Category

(Below is an explanation for each category on the questionnaire).

JOURNAL TITLE:	The correct current title of the journal.
EDITOR:	Name of editor responsible for manuscripts.
PHONE NO:	Telephone number of the editor.
MANUSCRIPT ADDRESS:	The correct address for submitting manuscripts.
TYPE OF ARTICLES:	The type of articles (research, review, theoretical, etc.) which the present editor indicates he or she will usually accept for publication.
MAJOR CONTENT AREAS:	The topics which are of prime interest to the the journal, as indicated by the journal's present editor.
NUMBER OF MANUSCRIPT COPIES:	The number of copies of a manuscript required by the editor for review purposes.
REVIEW PERIOD:	The editor's estimate of the average time interval between the time a manuscript is received and the time the author is notified whether it has been accepted or rejected.
PEER REVIEW:	Articles submitted for publication which are subject to "Peer Review", or reviewed by someone outside your editoral staff.
PUBLICATION LAG TIME:	The editor's estimate of the usual interval between the time an article is accepted for publication and the actual time it is published.
STYLE REQUIREMENTS:	The style requirements followed by the journal, i.e.: The Chicago Manual of Style; Council of Biology Editors Style Manual.
REVISED THESES AND DISSERTATIONS:	Whether MA/PhD graduate are encouraged to submit revised theses for publication.
CIRCULATION:	The size of the journal's circulation.
FREQUENCY:	The number of times per year the journal is published.
MISCELLANEOUS:	Any additional comments regarding your journal (s) that may be useful to potential authors.

ABBREVIATIONS

The following abbreviations were used consistently throughout the "Master List."

A	Annual
BA	Biannual
BM	Bimonthly
IR	Irregular
M	Monthly
PR	Peer Review
Q	Quarterly
SA	Semiannual
SM	Semimonthly
W	Weekly

Publishing Opportunities for Energy Research

MASTER LIST

The "Master List" contains descriptions of 524 periodical titles which have been determined to be likely targets for publishing energy-related research written in the English language. An attempt has been to include titles which an energy researcher might not normally consider. (See "Scope and Selection Methodology," page v for detail.) Titles are arranged alphabetically with titles beginning with an acronymn, such as the **AAPG BULLETIN,** preceding all other titles beginning with the letter "A."

1 **AAPG BULLETIN**

American Association of Petroleum Geologists
P.O. Box 979
Tulsa, OK 74101

Editor: Mrs. June Chronos, Managing Editor
Phone: 918-584-2555
ISSN: 0364-9849 **Circulation:** 44,300 **Frequency:** M
Style Sheet: Yes
Revised Theses or Diss.: Yes **Manuscript Copies:** 5
Review Period: 2 months (PR) **Lagtime:** 12 months
Types of Articles: Research and projects mostly in petroleum geology and related fields. Also accepts discussions and replies, book reviews, and memorials.
Major Contents: Petroleum geology.
Miscellaneous: Recent Bulletins, and "Instructions to Bulletin Authors" in every bulletin. (Index for year in December Bulletin.) Any thesis or dissertation must undergo stringent revision. "Elected Editor" for 1985 is Dr. James Helwig, Phone: 214-754-6257.

2 **AAPG EXPLORER**

American Association of Petroleum Geologists
P.O. BOX 979
Tulsa, OK 74101

Editor: Larry M. Nation
Phone: 918-584-2555
ISSN: 0195-2986 **Circulation:** 42,000 **Frequency:** M
Style Sheet: Yes
Revised Theses or Diss.: No **Manuscript Copies:** 1
Review Period: 1 month (PR) **Lagtime:** 1 month
Types of Articles: Research.
Major Contents: General and applied geology, geology of mineral deposits, historical and regional geology, paleozoology, paleobotany, general paleontology, and biostratigraphy.
Miscellaneous: Geology of petroleum exploration written in magazine style.

3 **A.G.A. MONTHLY**

American Gas Association
1515 Wilson Boulevard
Arlington, VA 22209

Editor: Athena S. Fristoe
Phone: 703-841-8686
ISSN: 0002-8584 **Circulation:** 6,500 **Frequency:** 11/year
Style Sheet: Yes
Revised Theses or Diss.: No **Manuscript Copies:** 2
Review Period: 2-3 months (PR) **Lagtime:** 4-6 months
Types of Articles: Research and projects, articles dealing with all areas of natural gas industry, particularly new technology applications holding potential for new or expanded markets for gas.
Major Contents: New technologies and markets such as cogeneration (gas-fired), biomass, natural gas-powered vehicles, gas air conditioning, and combined-cycle turbines. Also educational programs and public relations efforts.
Miscellaneous: This journal does not accept technical articles. Encourages items written in laymen's terms.

4 **AIAA JOURNAL**

American Institute of Aeronautics and Astronautics
1633 Broadway Street
New York, NY 10019

Editor: Dr. George W. Sutton, Helionetics, Inc., Laser Division, 5452 Oberlin Drive, San Diego, CA 92121
Phone: 212-581-4300
ISSN: 0001-1452 **Circulation:** 5,000 **Frequency:** M
Style Sheet: Yes
Revised Theses or Diss.: Yes **Manuscript Copies:** 4
Review Period: 3 months (PR) **Lagtime:** 6 months
Types of Articles: Research, reviews, and theoretical presentations.
Major Contents: This journal is devoted to the advancement of astronautics and aeronautics through the dissemination of original archival research papers disclosing new theoretical developments and/or experimental results. The topics include aeroacoustics, aerodynamics, combustion, fundamentals of propulsion, fluid mechanics and reacting flows, fundamental aspects of the aerospace environment, marine technology, lasers and associated phenomena, plasmas and magnetohydrodynamics, research instrumentation and facilities, structural mechanics and materials, and thermophysics and thermochemistry. Papers also are sought which review in an intensive manner the results of recent research developments on any of the topics listed above.
Miscellaneous: An official journal of the American Institute of Aeronautics and Astronautics.

5 **AICHE JOURNAL**

American Institute of Chemical Engineers
345 E. 47th Street
New York, NY 10017

Editor: Robert Kadlec, Editorial Offices, Chemical Engineering, Dow Building, University of Michigan, Ann Arbor, MI 48109
Phone: 313-763-5192
ISSN: 0001-1541 **Circulation:** 7,000 **Frequency:** M
Style Sheet: Yes
Revised Theses or Diss.: N/A **Manuscript Copies:** 3
Review Period: 6 months (PR) **Lagtime:** 6-9 months
Types of Articles: Research, reviews, R&D notes, and letters to the editor.
Major Contents: Significant theoretical and experimental developments in chemical engineering.

6 **ASHRAE JOURNAL**

Amer. Society, Heating, Refrigerating & Air Conditioning Engineer
1791 Tullie Circle, N.E.
Atlanta, GA 30329

Editor: Richard H. Green
Phone: 404-636-8400
ISSN: 0001-2491 **Circulation:** 42,000 **Frequency:** M
Style Sheet: Yes
Revised Theses or Diss.: Yes **Manuscript Copies:** 3
Review Period: 2 months **Lagtime:** 3-4 months
Types of Articles: Research, projects, reviews, and news.
Major Contents: Environmental control, heating, refrigeration, and air conditioning.

7 **ACORN**
Architectural Conservancy of Ontario, Inc.
86 Augusta Street
Port Hope, Ontario
Canada

Editor: Mrs. Marion W. Garland
Phone: 416-885-5769
ISSN: 0704-0083 **Circulation:** 2,000 **Frequency:** 3/year
Style Sheet: No
Revised Theses or Diss.: No **Manuscript Copies:** Varies
Review Period: Varies **Lagtime:** Varies
Types of Articles: News, project reviews, and conservancy reports.
Major Contents: Architectural conservancy.
Miscellaneous: Newsletter type articles pertaining to architectural conservation might be accepted. Not research oriented.

8 **ACTA ASTRONAUTICA**
Pergamon Press
Maxwell House, Fairview Park
Elmsford, NY 10523

Editor: Luigi G. Napolitano
Phone: 914-592-7700
ISSN: 0094-5765 **Circulation:** N/A **Frequency:** M
Style Sheet: Yes
Revised Theses or Diss.: N/A **Manuscript Copies:** 4
Review Period: 3 months (PR) **Lagtime:** 12 months
Types of Articles: Contributed papers, issues dedicated to IAF Congress and IAA symposia, and academy transactions.
Major Contents: Original contributions in all fields of basic engineering, life, and social space sciences and of space technology. Particular emphasis is given to the conception, design, and scientific exploration of space and the use of space for human welfare.

Miscellaneous: Contributions and notes are generally from the IAF or IAA membership or recommended by members. Acceptance rate is 90%. Formerly: ASTRONAUTICA ACTA, ISSN: 0004-6205.

9 ACTA GEODAETICA, GEOPHYSICA ET MONTANISTICA HUNGARICA

Institute of the Hungarian Academy of Sciences
Sopron
P.O.B. 5.H 9401
Hungary

Editor: A. Tarczy-Hornoch
Phone: 36-99-14-290
ISSN: 0001-5679 **Circulation:** N/A **Frequency:** Q
Style Sheet: Yes
Revised Theses or Diss.: Yes **Manuscript Copies:** 2
Review Period: 3 months **Lagtime:** 12 months
Types of Articles: Research papers and Hungarian and conference papers.

10 ACTA UNIVERSITATIS CAROLINAE-GEOLOGICA

Department of Petrology
Faculty of Science, Charles University
Albertov 6 128 43
Praha 2, Czechoslovakia

Editor: Professor J. Konta, Dr. Sc.
Phone: 297541
ISSN: 0001-7132 **Circulation:** 700 **Frequency:** Q
Style Sheet: No
Revised Theses or Diss.: No **Manuscript Copies:** 2
Review Period: 6 months **Lagtime:** 12 months
Types of Articles: Research and reviews.
Major Contents: Geology, mineralogy, petrology, mineral deposits, economic geology, paleontology, applied geophysics, and geochemistry.

11 ADVANCED BATTERY TECHNOLOGY

Robert Morey Association
P.O. Box 98
Dana Point, CA 92629

Editor: Robert Morey
Phone: 714-496-2574

ISSN: 0001-8627 **Circulation:** 1,500 **Frequency:** M
Style Sheet: No
Revised Theses or Diss.: No **Manuscript Copies:** 2
Review Period: 2 months (PR) **Lagtime:** 3-6 months
Types of Articles: Reports and news on developments in research, engineering, development, marketing, and production.
Major Contents: Research programs, new products, and new patents dealing with batteries, fuel cells, and associated electrochemistry.
Miscellaneous: Formerly: BATTERIES TODAY.

12 ADVANCES IN COLLOID AND INTERFACE SCIENCE

Elsevier Science Publishers B.V.
P.O. Box 211
1000 AE Amsterdam
The Netherlands

Editor: A.C. Zettlemover, Managing Editor, Provost and Vice President, 27 Alumni Building, Lehigh University, Bethelem, PA 18015
Phone: 215-861-3165
ISSN: 0001-8686 **Circulation:** N/A **Frequency:** 8/year
Style Sheet: Yes
Revised Theses or Diss.: No **Manuscript Copies:** N/A
Review Period: Varies (PR) **Lagtime:** 6 months
Types of Articles: Review articles of research.
Major Contents: Colloid and surface science, i.e., adsorption and pore structure; thermodynamic, electrical and optical properties of dispersed systems; nucleation processes; radiation scattering from particles; and rheology and phenomena through membranes.
Miscellaneous: Authors wishing to prepare a review for the journal should first send a brief proposal to any member of the editorial board or directly to the editor. Authors receive a modest per-page honorarium.

13 AERO SUN-TIMES

Alternative Energy Resource Organization
324 Fuller, C4
Helena, MT 59601

Editor: Wilbur Wood, P.O. Box 12, Roundup, MT 59072
Phone: 406-259-1958
ISSN: N/A **Circulation:** 1,000 **Frequency:** BM
Style Sheet: Yes
Revised Theses or Diss.: Yes **Manuscript Copies:** 1
Review Period: 2 months **Lagtime:** 1-6 months
Types of Articles: Projects, reviews of personal experiences with re-

newable energy systems or conservation techniques are preferred. Also, analyses, research and project reports, periodical and book reviews, occasional energy-environment-related fiction, and poetry.

Major Contents: Renewable energy and conservation, sustainable agriculture, locally based economic systems, bioregional reports, environmental problems, and/ or improvements, solutions.

Miscellaneous: Consumer oriented.

14 AGRICULTURAL AND FOREST METEOROLOGY

Elsevier Science Publishers B.V.
P.O. Box 211
1000 AE Amsterdam
The Netherlands

Editor: W.E. Reifsnyder, Editor-in-Chief, School of Forestry, Marsh Hall, 360 Prospect Street, New Haven, CT 06511

ISSN: 0168-1923 **Circulation:** N/A **Frequency:** SA

Style Sheet: No

Revised Theses or Diss.: N/A **Manuscript Copies:** 3

Review Period: Varies **Lagtime:** Varies

Types of Articles: Research, reviews, and contributions of an applied nature and basic scientific nature.

Major Contents: Original studies and comprehensive reviews in the interdisciplinary field of meteorology and climatology applied to agriculture and forestry.

15 AGRICULTURAL ENGINEERING

American Society of Agricultural Engineers Transactions
2950 Niles Road
P.O. Box 410
St. Joseph, MI 49085

Editor: James Basselman

Phone: 616-429-0300

ISSN: 0001-2351 **Circulation:** 2,400 **Frequency:** BM

Style Sheet: Yes

Revised Theses or Diss.: N/A **Manuscript Copies:** 5

Review Period: 1-3 months (PR) **Lagtime:** 9 months

Types of Articles: Research.

Major Contents: Various aspects of agricultural engineering including: power and machinery, soil and water, structures and environment, electric power and processing, food engineering, wind energy, solar energy, and food technology.

Miscellaneous: Acceptance rate is approximately 75%.

16 **ALABAMA SUNRISE**

c/o UAH-JEEC Public Service
University of Alabama in Huntsville
Huntsville, AL 35899

Editor: Bernard M. Levine, Director, Public Service Programs, The University of Alabama in Huntsville, Huntsville, AL 35899
Phone: 205-895-6257
ISSN: N/A **Circulation:** N/A **Frequency:** Q
Style Sheet: No
Revised Theses or Diss.: No **Manuscript Copies:** 1
Review Period: 3 weeks **Lagtime:** 6 months
Types of Articles: Surveys, projects, news, and reviews.
Major Contents: Solar energy technology, environment, biomass, passive solar heating, natural cooling, solar water heaters, photovoltaics, hydropower, and space heating.
Miscellaneous: Joint publication of the Alabama Sun Coalition and the Alabama Sun Energy Association. Formerly: SUNCLIPSE AND SUNSPOT.

17 **ALASKA CONSTRUCTION AND OIL**

Vernon Publications, Inc.
109 W. Mercer Street, C #19081
Seattle, WA 98119

Editor: Christine Laing
Phone: 206-285-2050
ISSN: 0002-4473 **Circulation:** 8,750 **Frequency:** M
Style Sheet: Yes
Revised Theses or Diss.: No **Manuscript Copies:** 1
Review Period: 1-2 months **Lagtime:** 2-6 months
Types of Articles: Research, projects, and reviews.
Major Contents: Alaska construction methods; construction, oil, and timber mining research; and economic analysis are all specific to Alaska.

18 **ALTERNATIVE ENERGY RETAILER**

70 Edwin Avenue
Box 2180
Waterbury, CO 06722

Editor: John Florian
Phone: 203-755-0158

ISSN: 0273-8163 **Circulation:** 13,000 **Frequency:** M
Style Sheet: Yes
Revised Theses or Diss.: No **Manuscript Copies:** 1
Review Period: 2 weeks **Lagtime:** 2-3 months
Types of Articles: Projects and reviews.
Major Contents: Solid fuel heating, solar and wind energy, and marketing solid fuel heating appliances.
Miscellaneous: Consumer oriented.

19 ALTERNATIVE SOURCES OF ENERGY MAGAZINE

Alternative Sources of Energy (ASE)
107 S. Central Avenue
Milaca, MN 56353

Editor: Larry Stoiaken
Phone: 612-983-6892
ISSN: 0146-1001 **Circulation:** 8,000 **Frequency:** BM
Style Sheet: Yes
Revised Theses or Diss.: No **Manuscript Copies:** 1
Review Period: 1 month **Lagtime:** 2 months
Types of Articles: Research, projects, and progress reports which present practical applications of energy technologies as a means of attaining some degree of energy independence.
Major Contents: Production of electricity from renewable energy-solar, wind, biomass, hydropower, and alcohol. Also reports on conservation, energy efficient construction, transportation, and photovoltaics.

20 AMBIO, A JOURNAL ON THE HUMAN ENVIRONMENT

Pergamon Press
Maxwell House, Fairview Park
Elmsford, NY 10523

Editor: D. Hinrichsen, The Royal Swedish Academy of Science, Box 50005, S-104 05 Stockhom, Sweden
Phone: 08-15-47-44
ISSN: 0044-7447 **Circulation:** 3,500 **Frequency:** BM
Style Sheet: Yes
Revised Theses or Diss.: Yes **Manuscript Copies:** 3
Review Period: 1-6 months (PR) **Lagtime:** 6-12 months
Types of Articles: Articles, research reports, and news.
Major Contents: Current environmental research dealing with all aspects of the environment. Energy/environmental topics relate to energy use, products, and by-products. Examples of such research are: coal and the environment, fuel/wood crisis, biomass alternatives, conservation

of natural resources, and industrial pollution.

Miscellaneous: An international journal dedicated to environmental research and management. Aims to present material in such a way as to be instructive to scientists in other fields, students, politicians, professional planners, and interested laymen.

21 **AMERICAN JOURNAL OF SCIENCE**

Yale University, Kline Geology Laboratory
Box 2161, Yale Station
New Haven, CT 06520

Editor: Marie C. Casey, Managing Editor, See Miscellaneous for "The Editors"

Phone: 203-432-4658

ISSN: 0002-9599 **Circulation:** 3,000 **Frequency:** 10/year

Style Sheet: See Miscellaneous

Revised Theses or Diss.: Yes **Manuscript Copies:** 4

Review Period: 2 months **Lagtime:** 12 months

Types of Articles: Research, projects, comments on previous communications.

Major Contents: Recognized as "The First Scientific Journal in the U.S. devoted to the geological sciences and related fields. Coverage is broad.

Miscellaneous: Editors: John Rodgers, John H. Ostrom and Robert Berner *Suggestions to Authors of the Reports of the U.S. Geological Survey.

22 **AMERICAN LABORATORY**

International Scientific Communication, Inc.
808 Kings Highway
P.O. Box 827
Fairfield, CT 06430

Editor: Dr. Brian Howard

Phone: 203-576-0500

ISSN: 0044-7749 **Circulation:** 130,000 **Frequency:** M

Style Sheet: Yes

Revised Theses or Diss.: Yes **Manuscript Copies:** 3

Review Period: 3 months **Lagtime:** 3-6 months

Types of Articles: Research, projects, and reviews.

Major Contents: Innovative use of chemical and biological instrumentation and equipment; thermal analysis/chromatography; laboratory automation/laboratory computers; spectroscopy; microscopy; chemical analysis; environmental analysis; and separation techniques.

23 **AMERICAN MINERALOGIST, THE**

Mineralogical Society of America
2000 Florida Avenue, N.W.
Washington, DC 20009

Editor: Michael J. Holdaway, The American Mineralogist, Department of Geological Sciences, Southern Methodist Universtiy, Dallas, TX 75275
Phone: 202-462-6713
ISSN: 0003-004X **Circulation:** 4,200 **Frequency:** BM
Style Sheet: Yes
Revised Theses or Diss.: Yes **Manuscript Copies:** 3
Review Period: 1-2 months (PR) **Lagtime:** 4-6 months
Types of Articles: Research.
Major Contents: Descriptive mineralogy, properties of minerals, experimental mineralogy and petrology, geochemistry, isotope mineralogy, mineralogical apparatus and techniques, mineral occurrences and deposits, paragenesis, petrography and petrogenesis, topographical mineralogy, and crystal chemistry, and crystallography.

24 **AMERICAN OIL AND GAS REPORTER, THE**

Domestic Petroleum Publishing, Inc.
1069 Parklane
Wichita, KS 67218

Editor: Bill Campbell
Phone: 316-681-3560
ISSN: 0145-9198 **Circulation:** 18,734 **Frequency:** M
Style Sheet: No
Revised Theses or Diss.: Yes **Manuscript Copies:** 1
Review Period: 2 weeks (PR) **Lagtime:** 1-2 months
Types of Articles: Research and projects.
Major Contents: Exploration, drilling, production and marketing of oil and natural gas. Legislative and regulatory topics affecting independent oil and gas producers. Drilling and production technology oriented toward financial ramifications rather than engineering specifications.
Miscellaneous: Serves 18 independent oil and gas producer associations as an official publication.

25 **AMERICAN SCIENTIST**

The Scientific Research Society
345 Whitney Avenue, Sigma Xi
New Haven, CT 06511

Editor: Michelle Press
Phone: 203-624-2566
ISSN: 0003-0996 **Circulation:** 125,000 **Frequency:** BM
Style Sheet: No
Revised Theses or Diss.: Yes **Manuscript Copies:** 3
Review Period: 2-3 months (PR) **Lagtime:** 3 months
Types of Articles: Research, projects, and reviews.
Major Contents: Recent scientific research encompassing all areas of science and providing enough background and interpretation to allow scientists in other fields to understand and appreciate the work.
Miscellaneous: Research findings usually should first appear in the primary journals.

26 AMERICAN WATER WORKS ASSOCIATION JOURNAL

American Water Works Association
6666 W. Quincy Avenue
Denver, CO 80235

Editor: Nancy M. Zeilig
Phone: 303-794-7711
ISSN: 0003-150X **Circulation:** 32,000 **Frequency:** M
Style Sheet: Yes
Revised Theses or Diss.: Yes **Manuscript Copies:** 2
Review Period: 3 months (PR) **Lagtime:** 6-12 months
Types of Articles: Research and technology, use histories, and project summaries.
Major Contents: Water quality, water resources, water utility management, bond issues, and distribution and network analysis.

27 ANALYTICAL CHEMISTRY

American Chemical Society
1155 16th Street, N.W.
Washington, DC 20036

Editor: George H. Morrison
Phone: 202-872-4570
ISSN: 0003-2700 **Circulation:** 36,000 **Frequency:** M
Style Sheet: Yes
Revised Theses or Diss.: Yes **Manuscript Copies:** 4
Review Period: 1-2 months (PR) **Lagtime:** 3 months
Types of Articles: Research, projects, and reviews.
Major Contents: Papers of interest to researchers working in all areas of analytical chemistry, both theoretical and technical.

28 ANNALES DE L'INSTITUT BELGE DU PETROLE

I.B.P. - Rue De La Science
4-1040 Brussels, Belgium

Editor: G.F. Van Hollebeke
Phone: 02-512-8767
ISSN: N/A **Circulation:** 1,000 **Frequency:** Q
Style Sheet: No
Revised Theses or Diss.: N/A **Manuscript Copies:** 2
Review Period: 6 months **Lagtime:** 6 months
Types of Articles: Research, projects, and general and specialized articles dealing with all areas of petroleum and petrochemistry.
Major Contents: Research of petroleum, exploitation of oil fields, transportation of crude oils and petroleum products, distribution, water and air pollution, energy conservation, and economical questions in relation with oil industry.

29 ANNALES DE LA SOCIETE GEOLOGIQUE DU NORD

Universite De Lille 1:Sciences De La Terre
59655 Villeneuve-Dascq Cedex, France

Editor: Mme P. Corsin
Phone: 20-G1-G2-22
ISSN: N/A **Circulation:** N/A **Frequency:** Q
Style Sheet: N/A
Revised Theses or Diss.: Yes **Manuscript Copies:** N/A
Review Period: Varies **Lagtime:** 12 months
Types of Articles: Research and reviews of books.
Major Contents: General and regional geology with a new orientation in northwest European regions around the north sea and applied geology (hydrogeology, coal geology, and petroleum geology.
Miscellaneous: Manuscripts may be submitted in French or English.

30 ANNALS OF BIOMEDICAL ENGINEERING

Pergamon Press
Maxwell House, Fairview Park
Elmsford, NY 10523

Editor: Dr. Hun H. Sun, Electrical Engineering, Biomedical Engineering & Science Institute, Drexel University, Philadelphia, PA 19104
Phone: 914-592-7700
ISSN: 0090-6964 **Circulation:** 2,200 **Frequency:** BM
Style Sheet: Yes

Revised Theses or Diss.: Yes **Manuscript Copies:** N/A
Review Period: Varies (PR) **Lagtime:** 6 months
Types of Articles: Reports, research, theses, abstracts, book reviews, patent disclosures, software surveys, and letters to editors.
Major Contents: Reports of original research on the application of sound engineering theory or practice to significant biomedical questions and problems, in the area of biomaterials, biomechanics and rehabilitation engineering, and instrumentation and physiological systems analysis.
Miscellaneous: Manuscripts should be sent to Professor Ernest O. Lang at editor's address. Incorporating: JOURNAL OF BIOENGINEERING, ISSN: 0145-3086.

31 **ANNALS OF NUCLEAR ENERGY**
Pergamon Press
Maxwell House, Fairview Park
Elmsford, NY 10523

Editor: Dr. L.W. Weaver, Dean of Engineering, 108 Ramsay Hall, Auburn University, Auburn, AL 36849
Phone: 914-592-7700
ISSN: 0306-4549 **Circulation:** 1,100 **Frequency:** M
Style Sheet: Yes
Revised Theses or Diss.: No **Manuscript Copies:** 3
Review Period: 1-2 months (PR) **Lagtime:** 3-4 months
Types of Articles: Research.
Major Contents: Current research and development in nuclear energy fission and fusion, including: nuclear reactors; thermonuclear power; safety; thermohydraulics; heat transfer and fluid dynamics; and dynamics and control.
Miscellaneous: Papers from authors outside the U.S. should be sent to the European Editor-Professor M.M.R. Williams, Nuclear Engineering Department, Queen Mary College, Mile End Road, London E1 4NS, England, Phone: 01-980-4611.

32 **ANNALS OF REGIONAL SCIENCE, THE**
Louisiana State University
3189 CEBA
Baton Rouge, LA 70803

Editor: Roger L. Burford
Phone: 504-388-2126
ISSN: 0570-1864 **Circulation:** 1,100 **Frequency:** 3/year
Style Sheet: Yes
Revised Theses or Diss.: Yes **Manuscript Copies:** 3

Review Period: 3 months (PR) **Lagtime:** 8-12 months
Types of Articles: Research and book reviews.
Major Contents: Interdisciplinary research and policy, particularly those having a regional or spatial dimension including: regional and urban economics; geography; sociology; politics; energy economics and policy; environmental economics and policy; migration; and housing.

33 APPLIANCE MANUFACTURER

Cahners Publishing Company
1350 E. Touhy Avenue, Box 5080
Des Plaines, IL 60018

Editor: Arthur Klein
Phone: 312-635-8800
ISSN: 0003-679X **Circulation:** 31,200 **Frequency:** M
Style Sheet: Yes
Revised Theses or Diss.: No **Manuscript Copies:** 1
Review Period: 1 month **Lagtime:** 1 month
Types of Articles: Projects and reviews.
Major Contents: Articles directly related to the design, manufacture and use of appliances, including: major appliances, electric and gas housewares, heating and air conditioning systems, and commercial and institutional appliances.
Miscellaneous: Trade oriented.

34 APPLIED AND ENVIRONMENTAL MICROBIOLOGY

American Society for Microbiology
1913 I Street, N.W.
Washington, DC 20006

Editor: James M. Tiedje
Phone: 202-833-9680
ISSN: 0099-2240 **Circulation:** 8,500 **Frequency:** M
Style Sheet: Yes
Revised Theses or Diss.: No **Manuscript Copies:** 2
Review Period: 1-2 months (PR) **Lagtime:** 2-3 months
Types of Articles: Original research.
Major Contents: All aspects of applied research as well as research of a genetic and molecular nature that focuses on topics of practical value and basic research in microbial ecology. Topics include microbiology in relationship to foods, and agriculture, industry, biotechnology, and public health; basic biological properties of bacteria, fungi, protozoa; and other simple eucaryotic organisms as related to microbial ecology.

35 **APPLIED ENERGY: AN INTERNATIONAL JOURNAL**

Applied Science Publishers
22 Rippleside Commercial Estate
Barking, Essey 1G11 0SA
England

Editor: Professor S.D. Probert, Centre for Thermal Insulation Studies, School of Mechanical Engineering, Cranfield Institute of Technology, Cranfield, Bedford, MK43 0AL, England
Phone: 0234-750111
ISSN: 0306-2619 **Circulation:** N/A **Frequency:** M
Style Sheet: Yes
Revised Theses or Diss.: No **Manuscript Copies:** 2
Review Period: 1-2 months (PR) **Lagtime:** 6 months
Types of Articles: Research, reviews, technical notes, short communications, book reviews, and reports on conferences and meetings.
Major Contents: Energy conversion, energy conservation, optimal use of energy and power resources, modeling and forecasting of energy needs and resources, conservation strategies, environmental and social impacts of energy policies and usage, combustion processes, ambient energy, and renewable energy sources.
Miscellaneous: This publication aims to bridge the gap between highly technical scientists practicing engineers, and decision-makers.

36 **APPLIED GEOGRAPHY**

Butterworth Scientific, Ltd.
P.O. Box 63, Westbury House
Bury Street
Guildford, Surrey GU2 5BH, England

Editor: David J. Briggs, Department of Geography, University of Sheffield, Sheffield S10 2TN, United Kingdom
ISSN: N/A **Circulation:** N/A **Frequency:** N/A
Style Sheet: N/A
Revised Theses or Diss.: N/A **Manuscript Copies:** N/A
Review Period: N/A **Lagtime:** N/A
Types of Articles: Research.
Major Contents: Topics include techniques, problems and results of resource survey and evaluation; environmental and social impacts, and implications of resource exploitation; principles, and policies and consequences of resource management and allocation.
Miscellaneous: Scope encompasses aspects of both human and physical geography, as well as parts of agriculture, ecology, planning and

politics.

37 **APPLIED OPTICS**

Optical Society of America, Inc.
American Institute of Physics
335 E. 45th Street
New York, NY 10017

Editor: J.N. Howard, Applied Optics, 7 Norman Road, Newton Highland, MA 02161
Phone: 617-332-1743
ISSN: 0003-6935 **Circulation:** 7,390 **Frequency:** SM
Style Sheet: AIP Style Manual.
Revised Theses or Diss.: N/A **Manuscript Copies:** 3
Review Period: N/A (PR) **Lagtime:** N/A
Types of Articles: Research, letters to the editor, patents, meetings calendar, and rapid communication.
Major Contents: The applications of facts, principles, and methods of optics. Papers of interest to applied physicists, space scientists, astronomers, information scientists, and electrical and mechanical engineers.
Miscellaneous: Manuscripts may be submitted in English, French, German, or Russian.

38 **APPLIED PHYSICS LETTERS**

American Institute of Physics
335 E. 45th Street
New York, NY 10017

Editor: Dr. Gilbert J. Perlow, Applied Physics Letters, Argonne National Laboratory, P.O. Box 296, Argonne, IL 60439
Phone: 312-972-4200
ISSN: 0003-6951 **Circulation:** 4,637 **Frequency:** SM
Style Sheet: AIP Style Manual.
Revised Theses or Diss.: Yes **Manuscript Copies:** 2
Review Period: 2-4 weeks (PR) **Lagtime:** 4 months
Types of Articles: Reports of original research of current importance.
Major Contents: Provides rapid dissemination of new developments in applied physics. Topics include: optics, lasers and laser applications, fluids, plasmas, and electrical discharges, structure mechanical thermal and electrical properties of condensed matter semiconductors, superconductors, acoustics, magnetism and magnetic materials, crystal growth, and other fields of applied physics.

Miscellaneous: Serves as a current awareness of new developments in the fields regularly covered by the Journal of Applied Physics. Revised theses and dissertations are accepted if less than 3 pages.

39 APPLIED PHYSICS:A SOLIDS AND SURFACES

Springer-Verlag Publishers
175 Fifth Avenue
New York, NY 10010

Editor: H.K.V. Lotsch, Springer-Verlag, P.O. Box 105280, D-6900 Heidelberg, Federal Republic of Germany
Phone: 212-460-1500
ISSN: 0721-7250 **Circulation:** N/A **Frequency:** M
Style Sheet: AIP Style Manual.
Revised Theses or Diss.: N/A **Manuscript Copies:** 3
Review Period: 1-2 months (PR) **Lagtime:** 2-4 months
Types of Articles: Research and letters of experimental and theoretical nature.
Major Contents: Investigations of phenomena, principles or methods of applied research on the condensed state including surface science and engineering such as: semiconductor physics, amorphous semiconductors, magnetism and superconductivity, metals and alloys, solid-state ionics, surface physics, chemisorption, ion implantation and sputtering, device physics, laser annealing and processing, integrated optics, fiber optics, and acoustic surface waves.
Miscellaneous: Formerly: APPLIED PHYSICS, ISSN: 0340-3793.

40 APPLIED PHYSICS:B PHOTOPHYSICS AND LASER CHEMISTRY

Springer-Verlag Publishers
175 Fifth Avenue
New York, NY 10010

Editor: H.K.V. Lotsch, Springer-Verlag, P.O. Box 105280, D-6900 Heidelberg 1, Federal Republic of Germany
Phone: 212-460-1500
ISSN: 0721-7269 **Circulation:** N/A **Frequency:** M
Style Sheet: AIP Style Manual.
Revised Theses or Diss.: N/A **Manuscript Copies:** 3
Review Period: 1-2 months (PR) **Lagtime:** 2-4 months
Types of Articles: Research, letters, and theoretical or experimental.
Major Contents: Investigations of phenomena, principles or methods of applied research of the gaseous state, including the application of laser radiation in chemistry. Examples include: high resolution laser spec-

troscopy, quantum electronics, ultra-fast phenomena, and chemical dynamics and structure.
Miscellaneous: Formerly: APPLIED PHYSICS, ISSN: 0340-3793.

41 **APPLIED SCIENTIFIC RESEARCH**
Martinus Nijhoff Publishers
Spuiboulevard 50
P.O. Box 163
3300 AD Dordrecht, The Netherlands

Editor: Professor R.B. Bird, Chemical Engineering Department, University of Wisconsin, 1415 Johnson Drive, Madison, WI 53706
Phone: 078-33-49-22
ISSN: 0003-6994 **Circulation:** N/A **Frequency:** Q
Style Sheet: Last issue of each volume.
Revised Theses or Diss.: Yes **Manuscript Copies:** 3
Review Period: 2 months (PR) **Lagtime:** 6 months
Types of Articles: Letters to the editor, original research, and review papers.
Major Contents: Thermal, mechanical, and electromagnetic phenomena in continua, i.e.: continuum mechanics; elasticity; plasticity and rheology; fluid dynamics; electromagnetic fields and waves; acoustics and elastodynamics; magnetohydrodynamics and electrohydrodynamics; heat and mass transfer; biological transport phenomena; and membrane physics.
Miscellaneous: Publication strives for international character and has a Board of Editors with world-wide representation. Authors from countries other than the U.S. should contact the main editorial office of the publisher.

42 **APPLIED SPECTROSCOPY REVIEWS**
Marcell Dekker, Inc.
270 Madison Avenue
New York, NY 10016

Editor: E.G. Brame
Phone: 302-798-2947
ISSN: 0570-4928 **Circulation:** 2,000 **Frequency:** Q
Style Sheet: Yes
Revised Theses or Diss.: No **Manuscript Copies:** 3
Review Period: 1-2 months **Lagtime:** 6-8 months
Types of Articles: Research and reviews.
Major Contents: Principles, methods, and applications of spectroscopy; discussions that relate physical concepts to chemical application, and

biochemical findings to geological understandings, etc.

Miscellaneous: This journal is international in scope with a board of editors from the major science oriented countries in the world. Contributions come from all these different countries. Most are invited but unsolicited contributions in spectroscopy are welcomed as well.

43 ARABIAN JOURNAL FOR SCIENCE AND ENGINEERING

University of Petroleum and Minerals
Box 8
Dhahran 31261,
Saudi Arabia

Editor: James Peeling
Phone: 03-860-5418
ISSN: 0377-9211 **Circulation:** 2,000 **Frequency:** Q
Style Sheet: Yes
Revised Theses or Diss.: No **Manuscript Copies:** 3
Review Period: 2 months (PR) **Lagtime:** 6 months

Types of Articles: Research results, research in progress, technical notes, state-of-the-art briefs, book reviews, and reader comments.

Major Contents: Considers for publication manuscripts in any scientific field, but focuses on chemistry, physics, geology, mathematics, and engineering.

44 ARCHITECTURAL SCIENCE REVIEW

University of Sydney
Department of Architectural Science
New South Wales 2006
Sidney, Australia

Editor: H.J. Cowan
Phone: 02-692-4801
ISSN: 0003-8628 **Circulation:** 900 **Frequency:** Q
Style Sheet: Yes
Revised Theses or Diss.: Yes **Manuscript Copies:** 1
Review Period: 2 months (PR) **Lagtime:** 6 months

Types of Articles: Research, projects, and reviews.

Major Contents: Architectural structures, materials, fire protection, acoustics and sound insulation, natural and artificial lighting and sunlight control, building and urban climatology, mechanical and electric equipment of buildings, computer applications to architectural design, and the relationship between science and architecture.

45 **ARCHITECTURAL TECHNOLOGY**

American Institute of Architects
1735 New York Avenue, N.W.
Washington, DC 20006

Editor: Mitch Ruda
Phone: 202-626-7590
ISSN: 0740-6142 **Circulation:** 46,421 **Frequency:** Q
Style Sheet: Yes
Revised Theses or Diss.: No **Manuscript Copies:** 1
Review Period: 1-2 months (PR) **Lagtime:** 2 months
Types of Articles: Research, projects, and reviews.
Major Contents: Topics in architectural technology, practice or design with emphasis on practical information, law, and management.
Miscellaneous: Preliminary calls or letters of inquiry are recommended.

46 **ARCHITECTURE: THE AIA JOURNAL**

The American Institute of Architects
1735 New York Avenue, N.W.
Washington, DC 20006

Editor: Donald Canty
Phone: 202-626-7300
ISSN: 0746-0554 **Circulation:** 43,000 **Frequency:** M
Style Sheet: Yes
Revised Theses or Diss.: Yes **Manuscript Copies:** 1
Review Period: 1-2 months (PR) **Lagtime:** 3-6 months
Types of Articles: Projects and research.
Major Contents: Architecture, design, evaluation of buildings, urban design, energy, preservation, and daylighting.

47 **ARCHIV FUER HYDROBIOLOGIE**

Limnolog Institute
Mainaustrasse 212
D-7750 Konstanz
Edd, West Germany

Editor: Professor Dr. H.J. Elster
Phone: 07-531-882-916
ISSN: 0003-9136 **Circulation:** N/A **Frequency:** M
Style Sheet: Yes
Revised Theses or Diss.: No **Manuscript Copies:** 2
Review Period: 1-3 months (PR) **Lagtime:** 12-16 months
Types of Articles: Research.

Major Contents: Entire field of limnology, fresh water research including treatment of problems of brackish and seawater in relationship to limnology.
Miscellaneous: Additional issues besides the main series: "Monographische Beitrage", "Advances in Limnology" and other supplement series.

48 **ASTROPHYSICAL LETTERS**
Gordon and Breach Science Publishers, Inc.
1 Park Avenue
New York, NY 10016

Editor: R.D. Henry, Astrophysical Letters, Physics Department, The Johns Hopkins University, Baltimore, MD 21218
Phone: 301-338-7351
ISSN: 0004-6388 **Circulation:** N/A **Frequency:** 3/year
Style Sheet: Yes
Revised Theses or Diss.: N/A **Manuscript Copies:** 3
Review Period: 1 month (PR) **Lagtime:** 6 months
Types of Articles: Research, communications, and book reviews.
Major Contents: Astrophysical research, both theoretical and observational.

49 **ATMOSPHERIC ENVIRONMENT**
Pergamon Press
Maxwell House, Fairview Park
Elmsford, NY 10523

Editor: James P. Lodge, Jr., 385 Broadway, Boulder, CO 80303
Phone: 303-449-7712
ISSN: 0004-6981 **Circulation:** 2,600 **Frequency:** M
Style Sheet: Yes
Revised Theses or Diss.: NA **Manuscript Copies:** 3
Review Period: 1 month (PR) **Lagtime:** 8 months
Types of Articles: Research, reviews, communications, technical notes, letters to the editors, and book reviews.
Major Contents: Aspects of human interactions with the atmospheric environment, including the administrative, economic, and political aspects of these interactions.
Miscellaneous: Papers should be of general, rather than local interest.

50 **ATOM**
United Kingdom Atomic Engineering Agency
11 Charles II Street
London SW1Y 4QP
England

Editor: Mrs. Wendy Peters
Phone: 01-930-5454
ISSN: 0004-7015 **Circulation:** 16,000 **Frequency:** M
Style Sheet: No
Revised Theses or Diss.: No **Manuscript Copies:** 1
Review Period: 1 month **Lagtime:** 1-2 months
Types of Articles: Projects and reviews.
Major Contents: All aspects of nuclear power including public acceptance and nuclear spinoff.
Miscellaneous: Articles should not exceed 3500 words.

51 **ATOMIC ENERGY LAW JOURNAL**
Invictus Publishing Corporation
180 S. Broadway
White Plains, NY 10605

Editor: Edward L. Weidenfeld
Phone: 914-949-6550
ISSN: 0004-7104 **Circulation:** 400 **Frequency:** Q
Style Sheet: A Uniform System of Citation.
Revised Theses or Diss.: Yes **Manuscript Copies:** N/A
Review Period: 3 months (PR) **Lagtime:** Varies
Types of Articles: Research, articles, and discussions of court cases decisions and regulations affecting atomic energy.
Major Contents: Developments and decisions concerning all aspects of atomic energy law and regulation on the local, national, and international level. Written for attorneys, executives, administrative agencies and researchers, and engineers in the field of atomic energy.

52 **ATOMKERNENERGIE/KERNTECHNIK**
DM 480. Verlag Karl Thiemi
Pilgersheimer Str. 38, 8000
Munich 90
West Germany

Editor: Professor Dr. A. Kraut, KARL THIEMIG VERLAG, Redaktion ATOMKERNENERGIE-KERNTECHNIK, Postfach 90 07 49, D-8000 Munchen, Fed., Republic of Germany
Phone: 089-62-48-226
ISSN: 0171-5747 **Circulation:** 1,000 **Frequency:** 8/year
Style Sheet: Yes
Revised Theses or Diss.: No **Manuscript Copies:** 3

Review Period: 2 months (PR) **Lagtime:** 3 months
Types of Articles: Original research papers, reviews, letters to the editor, and conference reports.
Major Contents: Mechanical engineering and technology; thermodynamics, fluidics and chemical engineering; electrical engineering and electronics; optical and radiation technology; energy and reactor technology; biology, and environment.
Miscellaneous: Information for authors reprinted in each issue. Manuscripts should be submitted in English or German.

53 **AUSTRALIAN GAS JOURNAL, THE**
P.O. Box 323
Canberra City, Act 2601, Australia

Editor: Anne Evans
ISSN: 0004-9166 **Circulation:** 2,500 **Frequency:** Q
Style Sheet: Yes
Revised Theses or Diss.: No **Manuscript Copies:** 1
Review Period: 1 month **Lagtime:** 3-4 months
Types of Articles: Projects and reviews.
Major Contents: All aspects of the gas industry.

54 **AUTOMATICA: THE IFAC JOURNAL**
Pergamon Press
Maxwell House, Fairview Park
Elmsford, NY 10523

Editor: George S. Alelby, Westinghouse Corporation, Box 746, MS-1297, Baltimore, MD 21203
Phone: 301-765-7768
ISSN: 0005-1098 **Circulation:** 2,000 **Frequency:** BM
Style Sheet: Yes
Revised Theses or Diss.: N/A **Manuscript Copies:** 6
Review Period: 3-4 months (PR) **Lagtime:** 9-10 months
Types of Articles: Research, technical surveys, and correspondences.
Major Contents: Theoretical and experimental research and original development work in the control of systems involving all facets of automatic control theory and its applications; modeling and computer control.
Miscellaneous: Work should emphasize applications of new results in analysis and synthesis with economic considerations given, where possible.

55 **BIO-JOULE NEWSLETTER**

Biomass Energy Institute, Inc.
1329 Niakwa Road
Winnipeg, Manitoba R2J 3T4
Canada

Editor: Peter Winter
Phone: 204-257-3891
ISSN: 0708-1936 **Circulation:** 2,500 **Frequency:** BM
Style Sheet: No
Revised Theses or Diss.: No **Manuscript Copies:** 1
Review Period: 3 months (PR) **Lagtime:** 3-5 months
Types of Articles: Research, projects, demonstrations, reviews, book reviews, news stories, interviews, and diary of events.
Major Contents: General and specialized articles dealing with all areas of renewable energy with particular emphases on bioenergy; biomass energy; alternative fuels; biotechnology; anaerobic digestion; alcohol production; energy-from-waste; recycling; photosynthesis; ethanol from biomass; methanol/ ethanol; energy outlook; supply and demand; solar and wind energy.

56 **BIOCYCLE: JOURNAL OF WASTE RECYCLING**

J.G. Press, Inc.
18 S. 7th Street, Box 351
Emmanus, PA 18049

Editor: Jerome Goldstein
Phone: 202-659-4102
ISSN: 0276-5055 **Circulation:** 8,000 **Frequency:** 8/year
Style Sheet: N/A
Revised Theses or Diss.: N/A **Manuscript Copies:** N/A
Review Period: N/A **Lagtime:** N/A
Types of Articles: Research, informative surveys and articles, policy analyses, and book reviews.
Major Contents: Focus is on biological management of industrial and municipal wastes, recycling, economic analysis of recycling applications, and resource recovery.
Miscellaneous: Formerly: COMPOST SCIENCE.

57 **BIOLOGICAL BULLETIN, THE**

Marine Biological Laboratory
Woods Hole, MA 02543

Editor: Dr. Charles B. Metz

Phone: 617-548-3705
ISSN: 0006-3185 **Circulation:** 3,100 **Frequency:** BM
Style Sheet: Yes
Revised Theses or Diss.: No **Manuscript Copies:** 3
Review Period: 1 month (PR) **Lagtime:** 4-6 months
Types of Articles: Research, projects, and reviews.
Major Contents: All aspects of biology; of general interest to biologists throughout the world.

58 **BIOMASS NEWS INTERNATIONAL**
254 Rue de Vaugirard
75740 Paris Cedex 15, France

Editor: Jean-Baptiste Prier
Phone: 532-27-19
ISSN: EN COURS **Circulation:** 1,000 **Frequency:** 11/year
Style Sheet: Yes
Revised Theses or Diss.: No **Manuscript Copies:** 1
Review Period: 1 month **Lagtime:** 1 month
Types of Articles: Research, projects, and reviews.
Major Contents: Biomass and its applications.

59 **BIOMASSE ACTUALITES**
254 Rue De Vaugirard
75740 Paris, Cedex 15, France

Editor: Pierre Benichou
Phone: 532-27-19
ISSN: 0294-3433 **Circulation:** 11,000 **Frequency:** 11/year
Style Sheet: Yes
Revised Theses or Diss.: No **Manuscript Copies:** 1
Review Period: 1 month **Lagtime:** 1 month
Types of Articles: Projects and reviews.
Major Contents: Utilization of biomass, biomass structure and chemical composition, biological conversions, engineering and economic aspects.

60 **BIOSCIENCE**
American Institute of Biological Science
1401 Wilson Boulevard
Arlington, VA 22209

Editor: William G. Van der Kloot
Phone: 703-527-6776

ISSN: 0006-3568 **Circulation:** 20,000 **Frequency:** 10/year
Style Sheet: See Miscellaneous
Revised Theses or Diss.: Yes **Manuscript Copies:** 3
Review Period: 3 months (PR) **Lagtime:** 6 months
Types of Articles: Research, projects, and reviews of non-technical nature.
Major Contents: A "Nonspecialist" periodical that focuses on the myriad advances being made in the biological, medical and agricultural sciences.
Miscellaneous: Council of Biology Editors Style Manual.

61 BOLLETTINO DI GEOFISICA. TEORICA ED APPLICATA
c/o OGS, P.O. Box 2011
34016 Trieste, Italy

Editor: I. Finetti
Phone: 040-214-0248
ISSN: 0006-6729 **Circulation:** N/A **Frequency:** Q
Style Sheet: Yes
Revised Theses or Diss.: No **Manuscript Copies:** 3
Review Period: 1-4 months (PR) **Lagtime:** 6 months
Types of Articles: Original scientific contributions, reviews, and letters.
Major Contents: Theoretical and experimental fields of solid earth geophysics; technologies and methodologies applied in geophysical studies; papers concerning physical oceanography and physics of the atmosphere with relevance to the interaction with the solid earth.
Miscellaneous: International in scope and acceptance. Contents, except for the title, are in English.

62 BOTANICAL REVIEW, THE
The New York Botanical Garden
Bronx, NY 10458

Editor: Dr. A. Cronquist
Phone: 212-220-8721
ISSN: 0006-8101 **Circulation:** 2,200 **Frequency:** Q
Style Sheet: Yes
Revised Theses or Diss.: Yes **Manuscript Copies:** 3
Review Period: 3 months (PR) **Lagtime:** 6-12 months
Types of Articles: Reviews only.
Major Contents: Botany.
Miscellaneous: Revised theses and dissertations are accepted only in the form of reviews.

63 BRIGHAM YOUNG UNIVERSITY GEOLOGY STUDIES
Brigham Young University Department of Geology Publication
258 Eyring Science Center
Provo, UT 84602

Editor: W. Kenneth Hamblin
Phone: 801-378-3918
ISSN: 0068 1016 **Circulation:** 600 **Frequency:** SA
Style Sheet: Geological Society Bulletin.
Revised Theses or Diss.: Yes **Manuscript Copies:** 4
Review Period: 3 months (PR) **Lagtime:** 6-8 months
Types of Articles: Research, projects, and reviews.
Major Contents: All aspects of geology, (western U.S.).

64 BRITISH POULTRY SCIENCE
Longman Group, Ltd.
Journals Division, Westgate House
The High
Harlow, Essex, England

Editor: B.O. Hughes, British Poultry Science, Poultry Research Centre, Roslin, Midlothian EH25 9PS, Scotland
Phone: 031-440-2726
ISSN: 0007-1668 **Circulation:** N/A **Frequency:** Q
Style Sheet: Yes
Revised Theses or Diss.: No **Manuscript Copies:** 2
Review Period: 1 month (PR) **Lagtime:** 10-14 months
Types of Articles: Research only.
Major Contents: Topics in poultry science with fundamental or applied value.

65 BUILDING AND ENVIRONMENT: THE INTERNATIONAL JOURNAL OF BUILDING SCIENCE AND ITS APPLICATIONS
Pergamon Press, Maxwell House
Fairview Park
Elmsford, NY 10523

Editor: Professor C.B. Wilson, University of Edinbrugh, 20 Chambers Street, Edinburgh EH1 1JZ, Scotland
Phone: 031-667-1011
ISSN: 0360-1323 **Circulation:** N/A **Frequency:** Q
Style Sheet: Yes
Revised Theses or Diss.: Yes **Manuscript Copies:** 2

Review Period: N/A (PR) **Lagtime:** 6 months
Types of Articles: Research, reviews, technical notes, letters to the editor, and book reviews.
Major Contents: Important topics in building and architectural research such as building research policies, problem identification and formulation, and applications in building design and construction.
Miscellaneous: Formerly: BUILDING SCIENCE.

66 **BULLETIN GEOLOGICAL SOCIETY OF MALAYSIA**

Geological Society of Malaysia
University of Malaya
Department of Geology
Kuala Lumpur 22-11, Malaysia

Editor: Dr. Teh Guan Hoe
Phone: 03-57-70-36
ISSN: 0126-6187 **Circulation:** 600 **Frequency:** A
Style Sheet: Yes
Revised Theses or Diss.: Yes **Manuscript Copies:** 2
Review Period: 6 months (PR) **Lagtime:** 6 months
Types of Articles: Research and projects.
Major Contents: Papers of general interest related to the geology, geoscience, marine areas, soils, or petroleum of southeast Asia.

67 **BULLETIN OF THE ATOMIC SCIENTISTS, THE**

Educational Foundation for Nuclear Science
5801 S. Kenwood Avenue
Chicago, IL 60637

Editor: Len Ackland
Phone: 312-363-5225
ISSN: 0096-3402 **Circulation:** 23,000 **Frequency:** 10/year
Style Sheet: University of Chicago Style Manual.
Revised Theses or Diss.: No **Manuscript Copies:** 2
Review Period: 1-2 months (PR) **Lagtime:** 2-3 months
Types of Articles: Non-technical articles and research.
Major Contents: Relationship of science to world affairs, especially questions of the arms race, superpower relations, and nuclear power.

68 **CIM BULLETIN**

Canadian Institute of Mining and Metallurgy
1130 Sherbrooke Street, W.
Suite 400
Montreal, Quebec H3A 2M8, Canada

Editor: Pierre Michaud
Phone: 514-842-3461
ISSN: 0317-0926 **Circulation:** 12,000 **Frequency:** M
Style Sheet: Yes
Revised Theses or Diss.: Yes **Manuscript Copies:** 3
Review Period: 6-18 months (PR) **Lagtime:** 6-18 months
Types of Articles: Research, projects, reviews, mining related special features, and membership activities.
Major Contents: All aspects of minerals industry: operational methods, specialized topics, computer applications, mineral economics, management methods, environmental control, health, and safety.

69 **CALIFORNIA GEOLOGY**
Division of Mines and Geology
1416 9th Street, Room 1341
Sacramento, CA 95814

Editor: Mary C. Woods, Editor-in-Chief
Phone: 916-445-0514
ISSN: 0026-4555 **Circulation:** 10,000 **Frequency:** M
Style Sheet: Yes
Revised Theses or Diss.: Yes **Manuscript Copies:** 1
Review Period: 1 month (PR) **Lagtime:** 2-3 months
Types of Articles: Research, reports, projects, and news items related to earth sciences in California.
Major Contents: Earth history, seismology, geophysics, gems and minerals, prospecting and mining, mineral resources, paleontology, California scenic resources, descriptions of geology of state parks, earth science items with public appeal, and geologic guides or field trip descriptions.
Miscellaneous: Articles are edited to conform to California Division of Mines and Geology Format.

70 **CANADIAN AERONAUTICS AND SPACE JOURNAL**
Canadian Aeronautics and Space Institute
75 Sparks Street
Saxe Building
Ottawa, Ontario K1P 5A5, Canada

Editor: A.A. Buchanan
Phone: 613-234-0919
ISSN: 0008-2821 **Circulation:** 2,350 **Frequency:** Q
Style Sheet: Yes

Revised Theses or Diss.: N/A **Manuscript Copies:** N/A
Review Period: 3 months (PR) **Lagtime:** 2-6 months
Types of Articles: Research, articles, and technical notes.
Major Contents: All aspects of aeronautics and space research.
Miscellaneous: Provides reprints if $65 page charge is paid.

71 CANADIAN AGRICULTURAL ENGINEERING

Canada Department of Agriculture
Research Station
Lethbridge
Alberta T1J 3B1, Canada

Editor: T.G. Sommerfeldt
Phone: 403-327-4561
ISSN: 0045-432X **Circulation:** 750 **Frequency:** SA
Style Sheet: Yes
Revised Theses or Diss.: No **Manuscript Copies:** 3
Review Period: 1-6 months (PR) **Lagtime:** 3-9 months
Types of Articles: Research and technical papers, general papers on education, research or extension, short technical notes on equipment development, and techniques or methods of analysis.
Major Contents: This journal covers a wide range of subject categories relating the application of engineering to agriculture, such as: soil and water; power and machinery; structures and environment; electrical power and processing.

72 CANADIAN CIVIL ENGINEER

EIC Publishers, Inc.
700 EIC Building
2050 Mansfield Street
Montreal, Quebec H3A 1Y9, Canada

Editor: Marta Meana
Phone: 514-842-8121
ISSN: 0825-7515 **Circulation:** 7,000 **Frequency:** BM
Style Sheet: No
Revised Theses or Diss.: No **Manuscript Copies:** 2
Review Period: 1-3 months (PR) **Lagtime:** 1-3 months
Types of Articles: Projects, reports, case studies, and general articles of a topical nature.
Major Contents: Environmental engineering, construction engineering, hydrotechnical engineering, computer application, and transportation.

73 CANADIAN ELECTRICAL ENGINEERING JOURNAL

EIC Publishers, Inc.
700 EIC Building
2050 Mansfield Street
Montreal, Quebec H3A 1Y9, Canada

Editor: Prof. V. Ramachandran, Department of Electrical Engineering, Concordia University, Montreal, Quebec H3G 1M8, Canada
Phone: 514-879-2828
ISSN: 0700-9216 **Circulation:** 1,600 **Frequency:** Q
Style Sheet: Yes
Revised Theses or Diss.: N/A **Manuscript Copies:** 5
Review Period: 3 months (PR) **Lagtime:** 6-8 months
Types of Articles: Research, technical, and reviews.
Major Contents: Electrical engineering; current electrical engineering practice; and the history of electrical engineering relevant to the Canadian scene.

74 CANADIAN GEOTECHNICAL JOURNAL

National Research Council of Canada
Public Information Branch
Ottawa, Ontario
K1A OR6, Canada

Editor: Dr. R.M. Quigley
Phone: 613-993-9101
ISSN: 0008-3674 **Circulation:** 2,750 **Frequency:** Q
Style Sheet: Yes
Revised Theses or Diss.: Yes **Manuscript Copies:** 3
Review Period: 3-6 months (PR) **Lagtime:** 6-12 months
Types of Articles: Reports of scientific work published as technical notes, discussions, and articles.
Major Contents: All aspects of geology.

75 CANADIAN JOURNAL OF BIOCHEMISTRY AND CELL BIOLOGY

Ramsay Wright Zoological Laboratory
25 Harbord Street
University of Toronto
Toronto, Ontario M5S 1A1, Canada

Editor: A.M. Zimmerman and K.G. Scrimgeour
Phone: 416-978-3509

ISSN: 0714-7511 **Circulation:** 2,430 **Frequency:** M
Style Sheet: Yes
Revised Theses or Diss.: Yes **Manuscript Copies:** 3
Review Period: 3 months (PR) **Lagtime:** 9 months
Types of Articles: Reports of original research in the form of full papers or communications; invited reviews and symposia.
Major Contents: Bioenergetics and metabolism, biomembranes and transport, cell structure and function, complex carbohydrates, enzymology, lipids, and proteins.

76 CANADIAN JOURNAL OF CHEMICAL ENGINEERING

Chemical Institute of Canada
151 Slater Street, Suite 906
Ottawa, Ontario
K1P 5H3, Canada

Editor: N. Epstein, Department of Chemical Engineering, University of British Columbia, Vancouver V6T 1W5, British Columbia
Phone: 613-233-5623
ISSN: 0008-4034 **Circulation:** 1,800 **Frequency:** BM
Style Sheet: No
Revised Theses or Diss.: Yes **Manuscript Copies:** N/A
Review Period: N/A (PR) **Lagtime:** N/A
Types of Articles: Research.
Major Contents: Chemical engineering science, industrial practice, and applied chemistry.
Miscellaneous: Formerly: CANADIAN JOURNAL OF TECHNOLOGY. Articles are written in French and English.

77 CANADIAN JOURNAL OF CHEMISTRY

National Research Council of Canada
Public Information Branch
Ottawa, Ontario
K1A OR6, Canada

Editor: W.A. Ayer, The Editorial Office, Canadian Journal of Chemistry, University of Alberta, Edmonton, Alberta T6 2G2, Canada
Phone: 613-993-0362
ISSN: 0008-4042 **Circulation:** 2,500 **Frequency:** BW
Style Sheet: N/A
Revised Theses or Diss.: N/A **Manuscript Copies:** N/A
Review Period: N/A (PR) **Lagtime:** N/A
Types of Articles: Original research.

Major Contents: All areas of chemistry including: general organic, solution properties, electrochemistry, thermochemistry, analytical, inorganic, bio-inorganic, organometallic, spectroscopy theoretical, nuclear magnetic resonance, physical organic, and the structure, synthesis, and bio-synthesis of natural products.
Miscellaneous: Authors are requested to submit manuscripts to one of ten subject editors listed in "Notice to Authors" section. This also may be obtained by writing to the Editorial Office (above). Manuscripts may be submitted in English or French.

78 CANADIAN JOURNAL OF CIVIL ENGINEERING
National Research Council of Canada
Public Information Branch
Ottawa, Ontaria
K1A OR6, Canada

Editor: C.H. Atkinson
Phone: 613-993-0428
ISSN: 0315-1468 **Circulation:** 5,500 **Frequency:** Q
Style Sheet: Yes
Revised Theses or Diss.: N/A **Manuscript Copies:** N/A
Review Period: 1-2 months (PR) **Lagtime:** 9-10 months
Types of Articles: Research, reviews, and book reviews.
Major Contents: All aspects of civil engineering.
Miscellaneous: Manuscripts may be submitted in English or French.

79 CANADIAN JOURNAL OF EARTH SCIENCES
National Research Council of Canada
Public Information Branch
Ottawa, Ontario
K1A OR6, Canada

Editor: W.G.E. Caldwell
Phone: 613-993-9101
ISSN: 0008-4077 **Circulation:** 4,650 **Frequency:** M
Style Sheet: Geowriting.
Revised Theses or Diss.: Yes **Manuscript Copies:** 3
Review Period: 3-6 months (PR) **Lagtime:** 6-12 months
Types of Articles: Research, notes, discussions, and short communications.
Major Contents: Earth sciences.

80 **CANADIAN JOURNAL OF MICROBIOLOGY**
National Research Council of Canada
Public Information Branch
Ottawa, Ontario
K1A OR6, Canada

Editor: K.G. Johnson
Phone: 613-993-9101
ISSN: 0008-4166 **Circulation:** 2,400 **Frequency:** M
Style Sheet: Yes
Revised Theses or Diss.: Yes **Manuscript Copies:** 3
Review Period: 3-6 months (PR) **Lagtime:** 6-12 months
Types of Articles: Articles, notes, and occasional reviews.
Major Contents: Topics in any field of microbiology which make original contributions to science. Includes research techniques/developments in instrumentation.

81 **CANADIAN JOURNAL OF PHYSICS**
National Research Council of Canada
Public Information Branch
Ottawa, Ontario
K1A OR6, Canada

Editor: G. Rostoker
Phone: 403-432-3465
ISSN: 0008-4204 **Circulation:** 1,650 **Frequency:** M
Style Sheet: Yes
Revised Theses or Diss.: Yes **Manuscript Copies:** 4
Review Period: 2-4 months (PR) **Lagtime:** 3-6 months
Types of Articles: Reports, articles, notes, and communications.
Major Contents: Reports representing a significant contribution to any area of physics.
Miscellaneous: Articles must be original research, though review articles may be considered.

82 **CANADIAN METALLURGICAL QUARTERLY**
Pergamon Press
Maxwell House, Fairview Park
Elmsford, NY 10523

Editor: Professor J.M. Toguri, University of Toronto, Canadian Metallurigcal Quarterly, Dept. of Metallurgy & Materials Science, Toronto, Ontario M5S 1A4, Canada
Phone: 914-592-7700

ISSN: 0008-4433 **Circulation:** 1,200 **Frequency:** Q
Style Sheet: Yes
Revised Theses or Diss.: No **Manuscript Copies:** 3
Review Period: 1 month **Lagtime:** 9 months
Types of Articles: Research and technical reviews.
Major Contents: Mineral processing, chemical and extraction metallurgy, physical metallurgy, and materials science.

83 **CANADIAN PETROLEUM**
Southam Business Publications, Ltd.
1450 Dons Mills Road
Don Mills, Ontario
M3B 2X7, Canada

Editor: Martin Keeley
Phone: 416-445-6641
ISSN: 0008-4735 **Circulation:** 10,770 **Frequency:** M
Style Sheet: No
Revised Theses or Diss.: No **Manuscript Copies:** 2
Review Period: 1-2 months **Lagtime:** 3-6 months
Types of Articles: Research and analytical papers.
Major Contents: Efficiencies in products and systems, market trends and statistics, and practical problem solving related to every aspect of Canada's oil and gas industry. Trade and information oriented. Accepts contributed articles but these relate to current awareness topics affecting the Canadian oil and gas industry. Policy papers and research news releases would be appropriate.

84 **CARBON**
Pergamon Press
Maxwell House, Fairview Park
Elmsford, NY 10523

Editor: Dr. Peter Thrawer, Department of Materials Science & Engineering, The Pennsylvania State University, University Park, PA 16802
Phone: 914-592-7700
ISSN: 0008-6223 **Circulation:** 1,600 **Frequency:** BM
Style Sheet: Yes
Revised Theses or Diss.: N/A **Manuscript Copies:** 3
Review Period: 2-4 months (PR) **Lagtime:** 4 months
Types of Articles: Original research, reviews, short communications, letters to the editor, and software surveys.

Major Contents: Physics, chemistry, and the scientific aspects of the technology of a class of materials ranging from organic crystals and polymers through chars and carbon to graphite. Only those organic substances which are related in some way to aromatic or tetrahedrally bonded carbonaceous solids, or which can be transformed into them by heat treatment or other means, are of interest to this Journal.

Miscellaneous: Articles published in CARBON may not be published elsewhere. Manuscripts may be submitted in English, German, or French, but German and French submissions must be accompanied by a short summary in English. English submissions are to include a 200 word English abstract.

85 CASOPIS PRO MINERALOGII A GEOLOGII

Geological Survey
Malostranske Nam. 19
118 21 Prague 1
Czechoslovakia

Editor: Jan Petranek
ISSN: 0008-7378 **Circulation:** 900 **Frequency:** Q
Style Sheet: Yes
Revised Theses or Diss.: Yes **Manuscript Copies:** 1
Review Period: 1-2 months **Lagtime:** 12 months
Types of Articles: Research, reviews, reports on the activities of the Czechoslovakia Society for Mineralogy and Geology, some personal reports (anniversaries, obituaries).
Major Contents: Geological sciences, especially paleontology, mineralogy, economic deposits, petrology, and geochemistry. Some regional geology.

86 CHEMICAL AND ENGINEERING NEWS

American Chemical Society
1155 16th Street, N.W.
Washington, DC 20036

Editor: Michael Heylin
Phone: 202-452-8923
ISSN: 0009-2347 **Circulation:** 135,000 **Frequency:** W
Style Sheet: Yes
Revised Theses or Diss.: Yes **Manuscript Copies:** N/A
Review Period: N/A (PR) **Lagtime:** N/A
Types of Articles: Research, projects, and reviews.
Major Contents: All aspects of current activity in the field of chemical engineering.

Miscellaneous: News oriented. A place for media coverage of a new process or finding, but not the place for technical articles. Except for brief news items, most articles are written by CHEMICAL AND ENGINEERING NEWS editors.

87 **CHEMICAL ENGINEERING**
McGraw-Hill Publishing Company
1221 Avenue of the Americas
New York, NY 10020

Editor: Nicholas P. Chopey
Phone: 212-997-3198
ISSN: 0009-2460 **Circulation:** 71,000 **Frequency:** BW
Style Sheet: Yes
Revised Theses or Diss.: No **Manuscript Copies:** 1
Review Period: 1-2 months (*PR) **Lagtime:** 6-12 months
Types of Articles: Reports and reviews.
Major Contents: The chemical process industries and chemical engineering technology, including: design, selection, and operation of commercially used equipment and processes; generic, practical-orientation articles related to industrial operations; engineering features, operation and maintenance, and materials.
Miscellaneous: Articles must contain practical information that can be applied to real problems. *Though not true "peer review" articles are reviewed by in-house experts. Honorarium is paid.

88 **CHEMICAL ENGINEERING PROGRESS**
American Institute of Chemical Engineers
345 E. 47th Street
New York, NY 10017

Editor: Agnes K. Dubber'y
Phone: 212-705-7154
ISSN: 0360-7275 **Circulation:** 55,000 **Frequency:** M
Style Sheet: Yes
Revised Theses or Diss.: No **Manuscript Copies:** 1
Review Period: 3-6 months (PR) **Lagtime:** 3-6 months
Types of Articles: Research and projects.
Major Contents: More specifically, research and papers addressing the following: pollution - biomass, air, water, flue gas desulfurization practices. Processes coal processing, shale oil and other synthetic fuels, ethanol from biomass, pulp and paper, alternate feedstocks for petrochemicals, sulfuric acid/phosphoric acid. Engineering practices safety and loss prevention, plant equipment, scrubbers, mixers, extractors,

pilot plant operations, separation, modular technology. Management/professionalism - economic, synfuels trends, salaries, career advancement, gasohol marketing, energy outlook, petrochemical markets, regulations, government-business-professional interrelationships. Supply/demand, engineering, and construction. Engineering technology; distillation; energy conservation; computers and process control; nuclear accident.

89 **CHEMIE DER ERDE**
M. 104 VEB Gustav Fischer Verlag
Villengang 2, Postfach 176,
DDR-6900 Jena
German Democratic Republic

Editor: Dr. Klaus Heide
ISSN: 0009-2819 **Circulation:** N/A **Frequency:** Q
Style Sheet: Yes
Revised Theses or Diss.: No **Manuscript Copies:** 2
Review Period: 1 month (PR) **Lagtime:** 6-8 months
Types of Articles: Research, projects, and book reviews.
Major Contents: Worldwide research on mineralogy, petrology, pedology, geochemistry, geology, limnology, cosmochemistry, and geography.
Miscellaneous: Manuscripts may be submitted in German or English. Author - supplied. Summaries are in both languages.

90 **CHILTON'S INSTRUMENTS AND CONTROL SYSTEMS**
Chilton's Company
Chilton Way
Radnor, PA 19089

Editor: Ron Kuhfeld, Managing Editor
Phone: 215-964-4000
ISSN: 0164-0089 **Circulation:** 80,000 **Frequency:** 13/YEAR
Style Sheet: Yes
Revised Theses or Diss.: Yes **Manuscript Copies:** 1
Review Period: 2 months **Lagtime:** 3-4 months
Types of Articles: Projects, reviews, and technical articles on state-of-the-art technology.
Major Contents: Instrumentation and control systems; industrial processing control systems; design shortcuts; engineering developments; solutions to existing problems; measuring and transmitting; controlling, computing and actuating.
Miscellaneous: A current awareness trade publication.

91 **CIVIL ENGINEERING**
American Society of Civil Engineers
345 E. 47th Street
New York, NY 10017

Editor: Virginia Fairweather
Phone: 212-705-7463
ISSN: 0360-0556 **Circulation:** 85,000 **Frequency:** M
Style Sheet: Yes
Revised Theses or Diss.: Yes **Manuscript Copies:** 1
Review Period: 1-2 months (PR) **Lagtime:** 6-14 months
Types of Articles: Research, projects, and reviews.
Major Contents: All aspects of civil engineering including professional and ethical subjects of broad application as well as current technical and professional activities.
Miscellaneous: Primary source of news to members of American Society of Civil Engineers.

92 **COAL MINING AND PROCESSING**
Maclean Hunter Publishing Corporation
300 W. Adams Street
Chicago, IL 60606

Editor: Eugene Guccione
Phone: 312-726-2802
ISSN: 0009-9961 **Circulation:** 22,000 **Frequency:** M
Style Sheet: Yes
Revised Theses or Diss.: Yes **Manuscript Copies:** 3
Review Period: 1 month **Lagtime:** 1-2 months
Types of Articles: Research, projects, and reviews.
Major Contents: All topics concerning the coal industry, including both surface and underground mining and the preparation and processing of coal.
Miscellaneous: While peer review is not incorporated, this is a viable avenue to announce new developments in coal technology.

93 **COASTAL ENGINEERING**
Elsevier Science Publishers B.V.
P.O. Box 330
1000 AH Amsterdam
The Netherlands

Editor: E. Byker

Phone: 020-5803-911
ISSN: 0378-3839 **Circulation:** N/A **Frequency:** Q
Style Sheet: First issue each year.
Revised Theses or Diss.: N/A **Manuscript Copies:** 3
Review Period: 1 month **Lagtime:** 6-9 months
Types of Articles: Research, reviews and case histories.
Major Contents: International medium for engineers working in the field of marine and coastal technology.

94 COLD REGIONS SCIENCE AND TECHNOLOGY

Elsevier Science Publishers
52 Vanderbilt Avenue
New York, NY 10017

Editor: M. Mellor
Phone: 212-867-9040
ISSN: 0165-232X **Circulation:** N/A **Frequency:** BM
Style Sheet: Yes
Revised Theses or Diss.: N/A **Manuscript Copies:** 3
Review Period: 2 weeks (PR) **Lagtime:** 6 months
Types of Articles: Research, technical, reviews, communications, letters, announcements, and book reviews.
Major Contents: Scientific and technical problems of cold environments, including natural and artificial environments. Problems relating to freezing of water, especially with the many forms of ice, snow, and frozen ground.
Miscellaneous: Primary language is English but papers in French and German are also accepted.

95 COLORADO SCHOOL OF MINES QUARTERLY

Colorado School of Mines Press Publications Center
Chauvenet Hall
Golden, CO 80401

Editor: Jon W. Raese
Phone: 303-273-3600
ISSN: 0163-9153 **Circulation:** 1,050 **Frequency:** Q
Style Sheet: Yes
Revised Theses or Diss.: Yes **Manuscript Copies:** 2
Review Period: 1-2 months (PR) **Lagtime:** 4 months
Types of Articles: Research.
Major Contents: Original unpublished studies dealing with mineral engineering, earth science, environment, and energy. Chemical and petroleum-refining engineering, chemistry and geochemistry, geology,

geophysics, metallurgy, mineral economics, petroleum, and mining. Energy, minerals, water, air, and ecology are acceptable if relevant to the understanding of the earth's resources.

96 COMBUSTION AND FLAME: THE JOURNAL OF THE COMBUSTION INSTITUTE

Elsevier Science Publishers
52 Vanderbilt Avenue
New York, NY 10017

Editor: Dr. Roger A. Strehlow, 105 Transportation Building, 104 S. Mathews Avenue, University of Illinois, Urbana, IL 61801
Phone: 217-333-3769
ISSN: 0010-2180 **Circulation:** 1,300 **Frequency:** M
Style Sheet: Yes
Revised Theses or Diss.: Yes **Manuscript Copies:** 4
Review Period: 1-2 months (PR) **Lagtime:** 5 months
Types of Articles: Research and critical reviews of research.
Major Contents: All aspects of combustion, including: experimental and theoretical investigations of combustion phenomena and closely allied matters; atmospheric pollution from combustion; combustion in practical systems; deflagration and detonation waves; electrical aspects of flames; spectroscopy and reaction kinetics.
Miscellaneous: No page charges. Reprints may be purchased.

97 COMBUSTION SCIENCE AND TECHNOLOGY

Gordon and Breach Science Publishers, Inc.
1 Park Avenue
New York, NY 10016

Editor: Dr. Irwin Glassman, Professor, Mechanical and Aerospace Engineering, School of Engineering and Applied Science, Princeton University, Princeton, NJ 08544
Phone: 212-689-0360
ISSN: 0010-2202 **Circulation:** N/A **Frequency:** Q
Style Sheet: Yes
Revised Theses or Diss.: N/A **Manuscript Copies:** 3
Review Period: 1 month (PR) **Lagtime:** 6-8 months
Types of Articles: Reviews, communications, and theoretical and experimental research.
Major Contents: Topics on the field of combustion, such as: flame and fire research; flame radiation; chemical fuels and propellants; reacting flows; thermochemistry; atmospheric chemistry and combustion phenomena related to aircraft gas turbines, automotive engines, furnaces,

and environmental studies.

98 COMMENTS ON ATOMIC AND MOLECULAR PHYSICS

Gordon and Breach Science Publishers, Ltd.
1 Bedford Street
London WC2E 9TP
England

Editor: Professor H.H. Stroke, Spectroscopy Laboratory, Department of Physics, New York University, 4 Washington Place, New York, NY 10003
Phone: 212-598-3204
ISSN: 0010-2687 **Circulation:** 700 **Frequency:** M
Style Sheet: AIP Style Manual.
Revised Theses or Diss.: No **Manuscript Copies:** 2
Review Period: 3 months (PR) **Lagtime:** 6 months
Types of Articles: Research trends, reviews with future projections, and conference reports.
Major Contents: Atomic physics; molecular physics; collisions; spectroscopy; atom - surface interactions; laser physics; and isotope shifts.
Miscellaneous: European editor is Frederik J. de Heer, FOM-INSTITUT, Postbus 41883, 1009 DB Amsterdam, The Netherlands, Phone: 020-946711.

99 COMMENTS ON PLASMA PHYSICS AND CONTROLLED FUSION

Gordon and Breach Science Publishers, Ltd.
1 Bedford Street
London WC2E 9TP
England

Editor: Burton D. Fried
Phone: 212-206-8900
ISSN: 0374-2806 **Circulation:** N/A **Frequency:** BM
Style Sheet: N/A
Revised Theses or Diss.: N/A **Manuscript Copies:** N/A
Review Period: N/A **Lagtime:** N/A
Types of Articles: Critical commentaries and book reviews.
Major Contents: Developments in plasma physics and controlled fusion relating to areas such as plasma lifetime energy loss, energy balance in fusion devices, neutral beam sources, microwave equipment, fusion power plant technology, and related topics.

100 COMPUTERS AND ELECTRICAL ENGINEERING

Pergamon Press
Maxwell House, Fairview Park
Elmsford, NY 10523

Editor: Dr. Harriet B. Rigas, Electrical and Computer Engineering Department, Naval Post Graduate School, Supt. Code 62, Monterey, CA 93943
Phone: 914-592-7700
ISSN: 0045-7906 **Circulation:** 1,200 **Frequency:** Q
Style Sheet: Yes
Revised Theses or Diss.: N/A **Manuscript Copies:** 2
Review Period: 3-6 months (PR) **Lagtime:** 12 months
Types of Articles: Research, applications, and tutorial papers.
Major Contents: The relationships between computer technology and electrical engineering, including: the applications of analog, hybrid, and digital computers; the technological issues associated with the design and development of new generation computers; and new and emerging aspects of computer technology.

101 COMPUTERS AND INDUSTRIAL ENGINEERING

Pergamon Press
Maxwell House, Fairview Park
Elmsford, NY 10523

Editor: Hamed K. Eldin, School of Industrial Engineering & Management, Engineering North Room 322, Oklahoma State University, Stillwater, OK 74078
Phone: 405-624-6055
ISSN: 0360-8352 **Circulation:** 1,000 **Frequency:** Q
Style Sheet: As in "Computer Languages."
Revised Theses or Diss.: N/A **Manuscript Copies:** 3
Review Period: 3 weeks (PR) **Lagtime:** 12 months
Types of Articles: Research, discussions, and comments on the journal's articles.
Major Contents: Aimed at the computerized industrial engineering applications, methodology for developing viable computer solutions to industrial engineering problems, as well as the implementation of different industrial engineering techniques on computers. Also accepted are papers on the state-of-the-art of computer applications in industrial engineering.

102 COMPUTERS AND OPERATIONS RESEARCH

Pergamon Press
Maxwell House, Fairview Park
Elmsford, NY 10523

Editor: Samuel J. Raff, 8312 Snug Hill Lane, Potomac, MD 20854
Phone: 914-592-7700
ISSN: 0305-0548 **Circulation:** 1,000 **Frequency:** Q
Style Sheet: Yes
Revised Theses or Diss.: N/A **Manuscript Copies:** 3
Review Period: 2 months **Lagtime:** 5 months
Types of Articles: Tutorials, state-of-the-art papers, research, and reviews.
Major Contents: The use of computers and operations research techniques to solve problems in any of the following fields: ecology; transportation; safety; reliability; urban planning; economics; inventory control; investment strategy; and military analysis.

103 COMPUTERS IN EXPLORATION AND DEVELOPMENT

American Association of Petroleum Geologists
P.O. Box 979
Tulsa, OK 74101

Editor: Edward A. Beaumont
Phone: 918-584-2555
ISSN: N/A **Circulation:** N/A **Frequency:** Q
Style Sheet: Yes
Revised Theses or Diss.: No **Manuscript Copies:** 4
Review Period: N/A (PR) **Lagtime:** N/A
Types of Articles: Book reviews, software reviews, hardware reviews, industry news and notes, techniques, and research.
Major Contents: Provides a forum for technical discussions on the use of computer technology in geology and related disciplines, and disseminates information on the application of computers—primarily in exploration and development geology—but including a healthy mix of contributions in reservoir engineering, computer basis, basin evaluation, well log analysis, geophysics, economics, geochemistry, and planimetrics, etc.
Miscellaneous: Begins regular publication early 1986.

104 CONNECTIONS

215 Pennsylvania Avenue, S.E.
Washington, DC 20003

Editor: Ted Jonas
Phone: 202-546-4996
ISSN: N/A **Circulation:** 1,500 **Frequency:** M
Style Sheet: No
Revised Theses or Diss.: No **Manuscript Copies:** 2
Review Period: 3 weeks **Lagtime:** 3-4 months
Types of Articles: Projects describing how states and local governments are achieving energy self-reliance through efficiency and renewable energy investments.
Major Contents: Least-cost energy programs, policies, projects; community-scaled activities and state energy planning. Comparative analyses on conventional energy sources versus efficiency and alternatives.

105 **CONSERVATION AND RECYCLING**
Pergamon Press
Maxwell House, Fairview Park
Elmsford, NY 10523

Editor: Michael B. Bever, Department of Materials Science & Engineering, Massachusetts Institute of Technology, Room 13-5026, Cambridge, MA 02139
Phone: 617-253-3323
ISSN: 0361-3658 **Circulation:** 1,035 **Frequency:** Q
Style Sheet: Yes
Revised Theses or Diss.: N/A **Manuscript Copies:** 2
Review Period: 1 month **Lagtime:** Varies
Types of Articles: Research, reviews, and titles and authors of doctoral theses in the fields of conservation and recycling.
Major Contents: The science and technology of conservation and reclamation of energy and materials, excluding air and water. Preferred topics include: the effect, use, and substitution of materials and the avoidance of their premature abandonment; recovery of materials for their original application; the use of materials previously considered to be waste; and the economics of energy attainable through recycling.

106 **CONTINENTAL SHELF RESEARCH**
School of Oceanography WB-10
University of Washington
Seattle, WA 98195

Editor: Professor R.W. Sternberg
Phone: 206-545-2960
ISSN: 0278-4343 **Circulation:** N/A **Frequency:** Q
Style Sheet: Yes

Revised Theses or Diss.: No **Manuscript Copies:** 4
Review Period: 1-2 months **Lagtime:** 5-6 months
Types of Articles: Research.
Major Contents: Physical oceanography, chemistry, ecology, and sedimentology dealing with the continental shelf.

107 CONTRIBUTIONS TO MINERALOGY AND PETROLOGY

Springer-Verlag Publishers
175 Fifth Avenue
New York, NY 10010

Editor: Dr. I.S.E. Carmichael, Executive Editor, Department of Geology and Geophysics, University of California, Berkeley, CA 94720
ISSN: 0010-7999 **Circulation:** 1,000 **Frequency:** M
Style Sheet: Yes
Revised Theses or Diss.: Yes **Manuscript Copies:** 2
Review Period: N/A (PR) **Lagtime:** N/A
Types of Articles: Research.
Major Contents: Contents include topics in geochemistry; isotope geology; the petrology and genesis of igneous, metamorphic and sedimentary rocks; experimental petrology and mineralogy; and distribution and significance of elements and their isotopes in rocks.

108 CONTROL AND COMPUTERS

Acta Press
P.O. Box 2481
Anaheim, CA 92804

Editor: M.H. Hamza, P.O. Box 25, Station G, Calgary, Alberta T3A 2G1, Canada
Phone: 403-270-3616
ISSN: 0730-9538 **Circulation:** N/A **Frequency:** 3/year
Style Sheet: No
Revised Theses or Diss.: Yes **Manuscript Copies:** 3
Review Period: 2 months (PR) **Lagtime:** N/A
Types of Articles: Research.
Major Contents: Control theory and its applications, with special emphasis on the area of computers in control, surveys and tutorials.

109 CRITICAL MASS ENERGY BULLETIN

Public Citizens, Inc.
215 Pennsylvania Avenue, S.E.
Washington, DC 20003

Editor: Whayne Dillehay
Phone: 202-546-4790
ISSN: N/A **Circulation:** 3,000 **Frequency:** M
Style Sheet: No
Revised Theses or Diss.: No **Manuscript Copies:** 2
Review Period: 3 weeks **Lagtime:** 3-4 months
Types of Articles: Comments addressing any aspect of energy. Features detailing energy issues, especially alternative energy sources.
Major Contents: All energy-related fields, including: nuclear power/nuclear weapons; alternate energy sources; and new energy technologies.
Miscellaneous: Formerly: CRITICAL MASS ENERGY JOURNAL.

110 **CURRENT**
Thomson Publications Pty., Ltd.
P.O. Box 8308
Johannesburg 2000
South Africa

Editor: Tony Fry
ISSN: N/A **Circulation:** 5,800 **Frequency:** 10/year
Style Sheet: Yes
Revised Theses or Diss.: No **Manuscript Copies:** 1
Review Period: 1 month **Lagtime:** 2 months
Types of Articles: Projects - reviews.
Major Contents: Topics in the electrical and electronic fields, mainly concerned with practical application rather than design or theory.

111 **DEFENSE ELECTRONICS**
E W Communications, Inc.
1170 E. Meadow Drive
Palo Alto, CA 94303

Editor: David M. Russell
Phone: 415-494-2800
ISSN: 0278-3479 **Circulation:** 39,000 **Frequency:** M
Style Sheet: Yes
Revised Theses or Diss.: No **Manuscript Copies:** 2
Review Period: 1 month (PR) **Lagtime:** 3-6 months
Types of Articles: Research, projects, reviews, project reviews, and technical/applications.
Major Contents: Electronic equipment and systems used by defense department. Articles should be concerned with applications, operations, and systems, as opposed to components.

112 DRILLING, THE WELLSITE PUBLICATION

Associated Publishers, Inc.
P.O. Box 7000
Dallas, TX 75209

Editor: Mitch Mayborn
Phone: 214-358-3455
ISSN: 0012-6241 **Circulation:** 29,000 **Frequency:** M
Style Sheet: Yes
Revised Theses or Diss.: Yes **Manuscript Copies:** 1
Review Period: 1 month **Lagtime:** 2-4 months
Types of Articles: Projects, reviews, and technical "How to" articles.
Major Contents: Field reports and equipment analysis dealing with drilling; well completion; well servicing and production. Topics cover the U.S.A. and international drilling, both land and offshore.

113 DYNAMICS OF ATMOSPHERES AND OCEANS

Elsevier Science Publishers B.V.
P.O. Box 211
1000 AE Amsterdam
The Netherlands

Editor: A.R. Robinson
Phone: 020-5803-911
ISSN: N/A **Circulation:** N/A **Frequency:** Q
Style Sheet: Yes
Revised Theses or Diss.: No **Manuscript Copies:** 2
Review Period: 1-3 months (PR) **Lagtime:** 10 months
Types of Articles: Research and reviews.
Major Contents: Fluid dynamics of planets; the dynamics of atmospheres and oceans and their interactions; related systems including planetary fluid interiors and stellar atmospheres; basic processes.

114 EOS TRANSACTION OF THE AMERICAN GEOPHYSICAL UNION

American Geophysical Union
2000 Florida Avenue, N.W.
Washington, DC 20009

Editor: A.F. Spilhaus, Jr.
Phone: 202-462-6903
ISSN: 0096-3941 **Circulation:** 15,200 **Frequency:** W
Style Sheet: Yes

Revised Theses or Diss.: Yes **Manuscript Copies:** 3
Review Period: 1 month (PR) **Lagtime:** 1 month
Types of Articles: Research, reports on scientific meetings, science news, book reviews, and in-house union affairs.
Major Contents: Science-related including earth science, meteorology, and oceanography.

115 **EPRI JOURNAL**

Electrical Power Research Institute
3412 Hillview Avenue, P.O. Box 10412
Palo Alto, CA 94303

Editor: Brent Barker, Editor-in-Chief
Phone: 415-855-2258
ISSN: 0362-3416 **Circulation:** 30,000 **Frequency:** 10/year
Style Sheet: Chicago Manual of Style.
Revised Theses or Diss.: No **Manuscript Copies:** 1
Review Period: 2 months (PR) **Lagtime:** 4-5 months
Types of Articles: Research.
Major Contents: All aspects of electric power research and development, and related environmental and economic fields.
Miscellaneous: Largely written in-house. Must be related to EPRI-sponsored research and development (R&D). Very few accepted.

116 **EARTH SHELTER LIVING**

Webco Publishing, Inc.
110 S. Greeley
Stillwater, MN 55082

Editor: Doug Seitz
Phone: 612-430-1113
ISSN: 0744-1932 **Circulation:** N/A **Frequency:** BM
Style Sheet: Yes
Revised Theses or Diss.: Yes **Manuscript Copies:** 1
Review Period: 2 weeks **Lagtime:** 3 months
Types of Articles: Slides and black-white photographs; plans; and articles about experiences with earth shelter construction.
Major Contents: Topics concerning earth sheltered structures or energy systems relating to earth sheltering, such as: construction of earth shelters; passive solar energy; and alternative life styles.
Miscellaneous: Consumer oriented.

117 **ECLOGAE GEOLOGICAE HELVETIAE**
ETH Zentrum
Versuchsanstalt fur Wasserbau (VAW)
CH-8092, Zurich
Switzerland

Editor: Dr. A. Lambert
Phone: ZURICH 256-4117
ISSN: 0012-9402 **Circulation:** 1,400 **Frequency:** 3/year
Style Sheet: Yes
Revised Theses or Diss.: Yes **Manuscript Copies:** 3
Review Period: 3 months (PR) **Lagtime:** 9 months
Types of Articles: Research, projects, and original scientific papers.
Major Contents: All aspects of Geology.
Miscellaneous: Manuscripts submitted by members of the Swiss Geological Society are given preference. However, space is also provided for papers by other authors.

118 **ECOLOGICAL MODELLING**
Elsevier Science Publishers B.V.
P.O. Box 330
1000 AH Amsterdam
The Netherlands

Editor: S.E. Jorgenson
Phone: 020-580-3911
ISSN: 0304-3800 **Circulation:** N/A **Frequency:** 16/year
Style Sheet: Yes
Revised Theses or Diss.: N/A **Manuscript Copies:** 3
Review Period: 3 months (PR) **Lagtime:** 12 months
Types of Articles: Research.
Major Contents: Combines mathematical models, system analysis, and computer techniques to show the impact of man on ecosystems.

119 **ECOLOGICAL MONOGRAPHS**
Ecological Society of America
Center for Environmental Studies
AZ State University
Tempe, AZ 85287

Editor: Lee N. Miller, Managing Editor, Ecological Society of America, Cornell University, Corson Hall, Ithaca, NY 14853
Phone: 607-256-3221

ISSN: 0012-9615 **Circulation:** 3,750 **Frequency:** Q
Style Sheet: See Miscellaneous
Revised Theses or Diss.: Yes **Manuscript Copies:** 4
Review Period: 3 months (PR) **Lagtime:** 10 months
Types of Articles: Theoretical and applied research. Discussions and essays that develop ecological theory.
Major Contents: Individuals, populations, and communities of organisms, in relation to the environment and ecosystem processes.
Miscellaneous: The Council of Biology Editors Style Manual.

120 **ECOLOGY**
Ecological Society of America
Center for Environmental Studies
Arizona State University
Tempe, AZ 85287

Editor: Lee N. Miller, Managing Editor, Ecological Society of America, Cornell University, Ithaca, NY 14853
Phone: 607-256-3221
ISSN: 0012-9658 **Circulation:** 8,600 **Frequency:** BM
Style Sheet: See Miscellaneous
Revised Theses or Diss.: Yes **Manuscript Copies:** 4
Review Period: 3 months (PR) **Lagtime:** 10 months
Types of Articles: Theoretical and applied research. Discussions and essays that develop ecological theory.
Major Contents: Individuals, populations, and communities of organisms, in relation to the environment, and ecosystem processes. All of the following as they relate to the individual organisms and their interaction: energy and time budgets, energy partitioning, energy balance, energy reserves, energy metabolism, energy content, energy allocation during growth. Also energetics of predation, competition and other species interactions; energy flow through various trophic levels; food chains and food webs.
Miscellaneous: The Council of Biology Editors Style Manual.

121 **ELECTRIC LETTER, THE**
Route 2
Mt. Vernon, IL 62864

Editor: John D. Turrel
Phone: 618-242-6549
ISSN: N/A **Circulation:** *1,000* **Frequency:** BW
Style Sheet: No
Revised Theses or Diss.: Yes **Manuscript Copies:** 1

Review Period: 2 weeks **Lagtime:** 1 month
Types of Articles: Research, projects, and reviews.
Major Contents: Electricity marketing, pricing, utilization, public communications, conservation, and load management.
Miscellaneous: This publication is a newsletter-type presentation.

122 **ELECTRIC LIGHT AND POWER**
Dun and Bradstreet Corporation
1301 S. Grove Avenue, Box 1030
Barrington, IL 60010

Editor: Robert A. Lincicome
Phone: 312-381-1840
ISSN: 0013-4120 **Circulation:** 19,715 **Frequency:** M
Style Sheet: No
Revised Theses or Diss.: No **Manuscript Copies:** 1
Review Period: 1 month **Lagtime:** 2-4 months
Types of Articles: Research and projects.
Major Contents: New developments for electric utility use; utility engineering applications; R&D in large power system technology. Electrical generation system design and operation; electric power transmission; and distribution technology.
Miscellaneous: All manuscripts are rewritten to meet journal requirements.

123 **ELECTRIC MACHINES AND POWER SYSTEMS**
Hemisphere Publishing Corporation
1010 Vermont Avenue, N.W.
Washington, DC 20005

Editor: Professor S.A. Nasar, Department of Electrical Engineering, University of Kentucky, 453 Anderson Hall, Lexington, Kentucky 40506-0046
Phone: 606-257-8042
ISSN: 0731-356X **Circulation:** 500 **Frequency:** BM
Style Sheet: Yes
Revised Theses or Diss.: Yes **Manuscript Copies:** 3
Review Period: 2 months (PR) **Lagtime:** 3-6 months
Types of Articles: Original research, reviews, theoretical-experimental, and design papers.
Major Contents: Electric machines, electromechanics, and power systems.

124 **ELECTRIC PERSPECTIVES**
Edison Electric Institute
1111 19th Street, N.W.
Washington, DC 20036

Editor: Jane Nunnelee
Phone: 202-828-7586
ISSN: 0364-474X **Circulation:** 11,000 **Frequency:** Q
Style Sheet: No
Revised Theses or Diss.: Yes **Manuscript Copies:** 1
Review Period: 2-3 months **Lagtime:** 3 months
Types of Articles: Research, projects, and reviews.
Major Contents: Topics relative to the reliable supply of electricity, including: regulation, technology, economics, marketing, management, and environment.

125 **ELECTRIC POWER SYSTEMS RESEARCH**
Elsevier Sequoia S.A.
P.O. Box 851
CH-1001 Lausanne 1
Switzerland

Editor: M.E. Council
Phone: 021-207-381
ISSN: 0378-7796 **Circulation:** N/A **Frequency:** Q
Style Sheet: Yes
Revised Theses or Diss.: Yes **Manuscript Copies:** 2
Review Period: 1-2 months (PR) **Lagtime:** 6-12 months
Types of Articles: Original research, applicational and design studies, comparative case studies, reviews, significant test data, survey articles, and guest editorials.
Major Contents: All aspects of electric power systems, including: generation, transmission, distribution and utilization of electric energy.

126 **ELECTRIC VEHICLE/BATTERY TECHNOLOGY NEWS**
6300 Hillcroft
Suite 603
Houston, TX 77081

Editor: Donald Saxman, 1511 Fifth Street, #126, Missouri City, TX 77489
Phone: 713-774-3942
ISSN: 0160-1059 **Circulation:** N/A **Frequency:** M
Style Sheet: No

Revised Theses or Diss.: No **Manuscript Copies:** 1
Review Period: 1 week **Lagtime:** 1 month
Types of Articles: Press releases, company announcements, and material for review.
Major Contents: Primary and secondary battery technology and industry; electric vehicles; and fuel cells.
Miscellaneous: Industry oriented.

127 **ELECTRICAL APPARATUS**
Barks Publications, Inc.
400 N. Michigan Avenue
Chicago, IL 60611

Editor: Elsie Dickson
Phone: 312-321-9440
ISSN: 0190-1370 **Circulation:** 15,000 **Frequency:** M
Style Sheet: Yes
Revised Theses or Diss.: Yes **Manuscript Copies:** 1
Review Period: 1 month **Lagtime:** 2 months
Types of Articles: Articles and reviews.
Major Contents: Reviews of electronics; repair, replacement and remanufacture of electromechanical and electronic systems for industry; developments in the total energy field; developments relating to electric motors, generators, transformers, and controls; energy efficiency; solar energy; and electric vehicles.
Miscellaneous: Unsolicited articles are rare; however, university-generated materials have acceptability not granted commercial sources.

128 **ELECTRICAL CONSULTANT (EC)**
Cleworth Publishing Company, Inc.
One River Road
Cos Cob, CT 06807

Editor: Robert B. Morgan
Phone: 203-661-5000
ISSN: 0361-4972 **Circulation:** 20,000 **Frequency:** BM
Style Sheet: Yes
Revised Theses or Diss.: No **Manuscript Copies:** 2
Review Period: 2 months **Lagtime:** 2 months
Types of Articles: Projects and reviews.
Major Contents: Electrical aspects and special features of recently designed projects, such as: interior electrical distribution; high-voltage electrical distribution; indoor/outdoor lighting; fire and security alarms; uninterruptible power supplies; cogeneration; and electric heating.

Miscellaneous: 95% articles are submitted. The audience is the professional electrical engineer who makes commercial, industrial, and residential engineering applications.

129 ELECTRICAL REVIEW INTERNATIONAL

Business Press International
Quadrant House, The Quadrant
Sutton, Surrey
SM2 5AS, England

Editor: T.J. Cogle
Phone: 01-661-3113
ISSN: 0013-4384 **Circulation:** 21,112 **Frequency:** W
Style Sheet: No
Revised Theses or Diss.: Yes **Manuscript Copies:** 1
Review Period: 1 month **Lagtime:** 1-3 months
Types of Articles: Projects, reviews, news stories, and features.
Major Contents: Industrial and commercial use of electricity; power generation, distribution and transmission; installation engineering, electrification, and power generation; renewable energy sources; training and education.

130 ELECTRICAL WORLD

McGraw-Hill Publishing Company
1221 Avenue of Americas
New York, NY 10020

Editor: William C. Hayes
Phone: 212-512-2000
ISSN: 0013-4457 **Circulation:** 32,500 **Frequency:** M
Style Sheet: Yes
Revised Theses or Diss.: No **Manuscript Copies:** 1
Review Period: 1-2 months **Lagtime:** 3 months
Types of Articles: Projects, reviews, and results of research in topic areas.
Major Contents: Generation; transmission; distribution; energy management; and load management.

131 ELECTROCHIMICA ACTA

University of Newcastle upon Tyne, School of Chemistry
Electrochemistry Research Laboratory
Newcastle upon Tyne
NE1 7RU, England

Editor: R.D. Armstrong, Editor-in-Chief
ISSN: 0013-4686 **Circulation:** 1,900 **Frequency:** M
Style Sheet: Yes
Revised Theses or Diss.: No **Manuscript Copies:** 3
Review Period: 3 weeks **Lagtime:** 6 months
Types of Articles: Research, short communications, and reviews.
Major Contents: Pure and applied electrochemistry.

132 ELECTROCOMPONENT SCIENCE AND TECHNOLOGY

Gordon and Breach Science Publishers, Ltd.
1 Bedford Street
London WC2E 9TP
England

Editor: D.S.Campbell
Phone: 01-836-5125
ISSN: N/A **Circulation:** N/A **Frequency:** Q
Revised Theses or Diss.: N/A **Manuscript Copies:** N/A
Review Period: N/A **Lagtime:** N/A
Types of Articles: Research.
Major Contents: Concerned with the science and technology of Passive electrical and electronic components.
Miscellaneous: Does accept unsolicited articles.

133 ELECTROMAGNETICS

Hemisphere Publishing Corporation
1010 Vermont Avenue, N.W.
Washington, DC 20005

Editor: Professor P.L.E. Uslenghi, College of Engineering, University of Illinois at Chicago, Box 4348, Chicago, IL 60680
Phone: 312-996-2311
ISSN: 0272-6343 **Circulation:** 500 **Frequency:** Q
Style Sheet: Yes
Revised Theses or Diss.: No **Manuscript Copies:** 3
Review Period: Varies (PR) **Lagtime:** 6-9 months
Types of Articles: Research, reviews, and theoretical-experimental.
Major Contents: Electromagnetic theory, antennas and redomes, arrays, high frequency techniques, extra low frequency, numerical techniques, scattering and diffraction, wave propagation, inverse scattering and remote sensing, plasmas and ionosphere, and geophysical applications.
Miscellaneous: Accepts articles based on theses or dissertations.

134 **ELECTRONIC NEWS**
Fairchild Publications, Inc.
7 E. 12th Street
New York, NY 10003

Editor: James Lydon
Phone: 212-741-4000
ISSN: 0013-4937 **Circulation:** 70,800 **Frequency:** W
Style Sheet: No
Revised Theses or Diss.: No **Manuscript Copies:** N/A
Review Period: 1 week **Lagtime:** 1 week
Types of Articles: News.
Major Contents: Covers electronics and computer markets including government, military, aerospace, industrial, commercial, consumer, and international news.
Miscellaneous: Journalistic in style and content. Accepts news of innovative research but does not detail the research.

135 **ELECTRONICS AND POWER**
Institution of Electrical Engineers
Savoy Place
London WC2R OBL
England

Editor: John Cooper
Phone: STD 0438-31-3311
ISSN: 0013-5127 **Circulation:** 53,211 **Frequency:** M
Style Sheet: Yes
Revised Theses or Diss.: No **Manuscript Copies:** 1
Review Period: 1 month **Lagtime:** 1-2 months
Types of Articles: Research, projects, and reviews.
Major Contents: All areas of electrical/electronics engineering.
Miscellaneous: Written for broad electrical/electronics engineering readership.

136 **ELECTRONICS INDUSTRY**
E S Publications, Ltd.
46150 Coombe Road
New Malden
Surrey KT3 4QH, England

Editor: Chris Palmer
Phone: 01-949-6621

ISSN: 0307-2401 **Circulation:** 21,000 **Frequency:** M
Style Sheet: Yes
Revised Theses or Diss.: No **Manuscript Copies:** 1
Review Period: 1-2 weeks **Lagtime:** 1-2 months
Types of Articles: Technical articles including circuit design. News concerning technological advances.
Major Contents: Electronics circuits; devices systems.
Miscellaneous: Preferably not too company oriented; a general overview is preferred.

137 **ELECTRONICS LETTERS**
Institutition of Electrical Engineers
P.O. Box 26
Hitchin, Herts.
SG5 1SA, England

Editor: C. Scales
ISSN: 0013-5194 **Circulation:** 2,200 **Frequency:** 25/year
Style Sheet: Yes
Revised Theses or Diss.: Yes **Manuscript Copies:** 4
Review Period: 6 months (PR) **Lagtime:** 6 months
Types of Articles: Short papers in letter form.
Major Contents: Short papers in letter form presenting research in all aspects of electronics, especially advances being made in optical communications.

138 **ELECTRONICS TODAY**
Moorshead Publications, Ltd.
25 Overlea Boulevard
Unit 4, Toronto
Ontario, M4H 1B1, Canada

Editor: William Markwick
Phone: 416-423-3262
ISSN: 0703-8984 **Circulation:** 20,000 **Frequency:** M
Style Sheet: Yes
Revised Theses or Diss.: No **Manuscript Copies:** 1
Review Period: 1 month **Lagtime:** 2 months
Types of Articles: Articles, projects, and product reviews.
Major Contents: Electronics and electronic related developments, including: electronics research, communications, and medical and military applications.
Miscellaneous: Projects should be fully documented with accompanying diagrams and/or photographs (black/white) of good quality. Articles may

be submitted in Wordstar on Apple DOS files or an Apple CP/M floppy disk.

139 **ENERGIA**
Ingenieria Quimica S.A., Triana 51
Madrid 16, Spain

Editor: Fernando Vazquez
Phone: Madrid 457-64-00
ISSN: 0210-2056 **Circulation:** 5,000 **Frequency:** BM
Style Sheet: No
Revised Theses or Diss.: Yes **Manuscript Copies:** 1
Review Period: 2 months **Lagtime:** 2 months
Types of Articles: Research, projects, and reviews.
Major Contents: Electric power related topics such as: policy and planning; generation; transportation and distribution; design; construction and operation of power generation plants.

140 **ENERGIA NUCLEAR**
Ciudad Universitaria, Avda. Complutense
Junta De Energia Nuclear
Madrid 3, Spain
ISSN: 0013-7324 **Circulation:** 1,200 **Frequency:** BM
Style Sheet: Yes
Revised Theses or Diss.: Yes **Manuscript Copies:** 2
Review Period: 3 months **Lagtime:** 2-6 months
Types of Articles: Reviews, research, technology, and projects.
Major Contents: Peaceful uses of atomic energy.

141 **ENERGIA NUCLEAR E AGRICULTURA**
Cena - Library
P.O. Box 96
13.400
Piracicaba, SP, Brasil

Editor: Board of Editors
ISSN: 0100-3593 **Circulation:** 300 **Frequency:** SA
Style Sheet: No
Revised Theses or Diss.: Yes **Manuscript Copies:** 2
Review Period: 3-4 months **Lagtime:** 6 months
Types of Articles: Research, reviews, and technical notes.
Major Contents: Nuclear techniques in agriculture, biology, and ecology.

142 **ENERGY**

Pergamon Press
Maxwell House, Fairview Park
Elmsford, NY 10523

Editor: Professor S.S. Penner, Energy Center, University of California at San Diego, La Jolla, CA 92093
Phone: 714-452-4284
ISSN: 0360-5442 **Circulation:** 1,100 **Frequency:** M
Style Sheet: Inside back cover of journal.
Revised Theses or Diss.: Yes **Manuscript Copies:** 2
Review Period: 1 month (PR) **Lagtime:** 6 months
Types of Articles: Research, original papers, software survey, projects, and reviews.
Major Contents: Multidisciplinary focus for activities relating to the development, economics, assessment and management of energy-related programs.
Miscellaneous: International in scope.

143 **ENERGY AND BUILDINGS**

Elsevier Sequoia S.A.
P.O. Box 851
CH-1001 Lausanne 1
Switzerland

Editor: Stephen Selkowitz, Executive Editor, Building 90, Room 3111, Lawrence Berkeley Laboratory, Berkeley, CA 94720
Phone: 415-486-5111
ISSN: 0378-7788 **Circulation:** N/A **Frequency:** Q
Style Sheet: Yes
Revised Theses or Diss.: Yes **Manuscript Copies:** 3
Review Period: N/A **Lagtime:** N/A
Types of Articles: Original research, reviews, technical communications, letters to the editor, reviews of books and reports, and notices of meetings and conventions.
Major Contents: All topics relating to energy efficiency in the built environment.
Miscellaneous: Submit articles to main office in Switzerland.

144 **ENERGY AND HOUSING REPORT: THE NATIONAL NEWSLETTER ON RESIDENTIAL ENERGY CONSERVATION**

Business Publishing, Inc.
951 Pershing Drive
Silver Spring, MD 20910

Editor: Allan L. Frank
Phone: 301-587-6300
ISSN: 0731-6313 **Circulation:** N/A **Frequency:** M
Style Sheet: Yes
Revised Theses or Diss.: No **Manuscript Copies:** 1
Review Period: 1-2 months **Lagtime:** 3 months
Types of Articles: Projects and reviews.
Major Contents: Residential energy management and conservation.
Miscellaneous: News articles are accepted, but author credits are not given.

145 ENERGY AND MINERALS RESOURCES: THE WORLDWIDE REPORT ON INTEGRATED NATURAL RESOURCES MANAGEMENT

Business Publishers, Inc.
951 Pershing Drive
Silver Spring, MD 20910

Editor: Pat Maio
Phone: 301-587-6300
ISSN: 0278-5099 **Circulation:** N/A **Frequency:** W
Style Sheet: Yes
Revised Theses or Diss.: No **Manuscript Copies:** 1
Review Period: 1-2 weeks **Lagtime:** 1 month
Types of Articles: Research, projects, and reviews.
Major Contents: All aspects of national and international development in energy and minerals resources; fuel and nonfuel minerals, including: coal, synfuels, oil, and natural gas.
Miscellaneous: News articles are accepted, but authors credits are not given.

146 ENERGY CONVERSION AND MANAGEMENT

Pergamon Press
Maxwell House, Fairview Park
Elmsford, NY 10523

Editor: Dr. Jesse Denton, Energy Conversion and Management, P.O. Box 355, Belton, TX 76513
Phone: 914-592-7700
ISSN: 0196-8904 **Circulation:** 1,300 **Frequency:** Q
Style Sheet: Yes
Revised Theses or Diss.: N/A **Manuscript Copies:** N/A
Review Period: 3 months (PR) **Lagtime:** 6 months
Types of Articles: Research, papers, and reviews.

Major Contents: Energy conversion, fuel conservation, and energy management. All sources, forms, and uses of energy and conversion processes are appropriate.

147 **ENERGY DEVELOPMENTS IN JAPAN**
Rumford Publishing Company
Box 5370
Chicago, IL 60680

Editor: Y. Mori, Tokyo Institute of Technology, 12-1, Ohokayama, 2-Chome, Meguro-Ku, Tokyo 152, Japan
ISSN: 0161-8091 **Circulation:** N/A **Frequency:** Q
Style Sheet: Yes
Revised Theses or Diss.: No **Manuscript Copies:** N/A
Review Period: N/A **Lagtime:** N/A
Types of Articles: Research, development, and reviews.
Major Contents: All aspects of systemology and use of energy in Japanese factories and plants with particular emphasis on the latest methods employed for the increased efficiency of energy production and conservation in Japanese industry.
Miscellaneous: Papers from authors in the U.S. should be sent to J. P. Hartnett, Energy Resources Center, University of Illinois, P.O. Box 4348, Chicago, IL 60680, Phone: 312-996-4490.

148 **ENERGY ECONOMICS**
Butterworth Scientific, Ltd.
P.O. Box 63, Westbury House
Bury Street
Guildford, Surrey GU2 5BH, England

Editor: Dr. Homa Motamen, Imperial College of Science & Technology, University of London, 53 Prince's Gate, London SW7 2PG, United Kingdom
ISSN: 0140-9883 **Circulation:** N/A **Frequency:** BM
Style Sheet: Yes
Revised Theses or Diss.: Yes **Manuscript Copies:** 3
Review Period: 1-3 months (PR) **Lagtime:** 6-12 months
Types of Articles: Research , reviews, theory papers, book reviews, and rejoinders.
Major Contents: Economic theory and its application to all forms of energy, methodology, statistics and mathematical modeling; all aspects of economic analysis of energy issues - pricing, tariff structures, depletion, elasticities, energy/GDP relationships, investment appraisal, inflation effects, sector substitution, conservation, international trade effects,

shadow prices, macroeconomics, demand/supply studies.
Miscellaneous: The journal also contains a list of recent papers published in the field of energy economics in other periodicals or working papers. Written for the professional economist and those concerned with the economic analysis of energy issues.

149 ENERGY ECONOMICS, POLICY AND MANAGEMENT

Fairmont Press
P.O. Box 14227
Atlanta, GA 30324

Editor: William F. Payne, Star Route, Marble Hill, GA 30148
Phone: 404-447-5314
ISSN: 0275-7966 **Circulation:** 5,453 **Frequency:** Q
Style Sheet: McGraw Hill Style Manual.
Revised Theses or Diss.: Yes **Manuscript Copies:** 2
Review Period: 1 month **Lagtime:** 4 months
Types of Articles: Pragmatic, not academic, projects, and reviews.
Major Contents: Energy economics, energy-finance, energy forecasting, international energy, and energy politics.
Miscellaneous: Though "not academic," the editor states that the article must be interesting, topical, original, sophisticated, professional. Maximum length 18 double-spaced typewritten pages and preferably 12-15 pages.

150 ENERGY ENGINEERING

Fairmont Press
G.P.O. Box 411
Brooklyn, NY 11201

Editor: Richard L. Koral
Phone: 404-447-5314
ISSN: 0199-8595 **Circulation:** 7,000 **Frequency:** BM
Style Sheet: No
Revised Theses or Diss.: Yes **Manuscript Copies:** 1
Review Period: 1-2 months (PR) **Lagtime:** 6 months
Types of Articles: Research, design project, state-of-the-art-reviews, and letters from point of view engineering or architectural designer.
Major Contents: New energy sources, alternate energy (cogeneration, solar, wind), energy conservation, and energy auditing.

151 ENERGY EXPLORATION AND EXPLOITATION

Elsevier Applied Science Publications, Ltd.
Crown House, Linton Road
Barking
Essex 1G11 8JU, England

Editor: D.C. Ion, R. McQuillin, and D. Abbot
Phone: 01-594-7272
ISSN: 0144-5987 **Circulation:** N/A **Frequency:** Q
Style Sheet: Yes
Revised Theses or Diss.: No **Manuscript Copies:** 3
Review Period: 1-2 months (PR) **Lagtime:** 3 months
Types of Articles: Original articles, reviews of major energy topics, conference reports, book reviews, research reports, and project histories.
Major Contents: Exploration for and exploitation of the world's energy resources; advances in geological, geophysical and engineering techniques in the exploration, prospect assessment, exploitation and conversion of fossil fuels as well as uranium; prospects for exploiting available resources; economic projections and assessments; case histories; conventional sources; environmental assessments of impact of exploration; and exploration and use of various energy resources.

152 **ENERGY IN AGRICULTURE**

Elsevier Science Publishers B.V.
P.O. Box 211
1000 AE Amsterdam
The Netherlands

Editor: B.A. Stout
ISSN: 0167-5826 **Circulation:** 800 **Frequency:** Q
Style Sheet: N/A
Revised Theses or Diss.: N/A **Manuscript Copies:** N/A
Review Period: N/A **Lagtime:** N/A
Types of Articles: Research.
Major Contents: Topics include: energy and food systems, energy use and conservation in agricultural production, and energy production from biomass.

153 **ENERGY JOURNAL**

Oelgeschlager, Gunn, and Hain Publishers, Inc.
131 Clarendon Street
Boston, MA 02116

Editor: Helmut J. Frank, Department of Economics, University of Arizona, Tucson, AZ 85721
Phone: 202-293-5913
ISSN: 0195-6574 **Circulation:** 1,000 **Frequency:** Q
Style Sheet: Chicago Manual of Style.

Revised Theses or Diss.: Yes **Manuscript Copies:** 3
Review Period: 2-4 months (PR) **Lagtime:** 6-9 months
Types of Articles: Technical notes, book reviews, letters to the editor, research, notes, and comments.
Major Contents: Energy related topics with interdisciplinary and international coverage, emphasizing economics, planning, and policy. Subjects cover the many facets of oil and gas exports and imports, utility financing, energy models, alternative energy sources, energy product tariffs, and regulations.
Miscellaneous: A publication of the Energy Economics Educational Foundation in association with the International Association of Energy Economists.

154 **ENERGY LAW JOURNAL**
Federal Energy Bar Association
888 16th Street, N.W.
Washington, DC 20006

Editor: William A. Mogel. Editor-in-Chief
Phone: 202-822-8888
ISSN: 0270-9163 **Circulation:** 1,750 **Frequency:** SA
Style Sheet: A Uniform System of Citation.
Revised Theses or Diss.: Yes **Manuscript Copies:** 1
Review Period: 2-3 months (PR) **Lagtime:** 6 months
Types of Articles: Political research, policy reviews, book reviews, committee reports, case decisions, and discussions.
Major Contents: Federal energy law relating to the production, development, conservation, transmission, and economic regulation of energy.
Miscellaneous: Formerly: JOURNAL OF CONTEMPORARY LAW.

155 **ENERGY MANAGEMENT TECHNOLOGY**
Walker-Davis Publications, Inc.
2500 Office Center
Willow Grove, PA 19090

Editor: Frank McGill
Phone: 215-657-3203
ISSN: 0745-984X **Circulation:** 50,000 **Frequency:** 9/year
Style Sheet: Yes
Revised Theses or Diss.: No **Manuscript Copies:** 1
Review Period: 1-3 weeks **Lagtime:** 2-6 months
Types of Articles: Energy casebooks; technology updates; energy engineering articles; energy worksheets; and interviews "special report" section.

Major Contents: EMS and other facilities management systems, HVAC controls and systems, energy measuring and auditing systems, plus a wide range of specific topics related to energy control (infrared detection, motors and controls, boiler and combustion controls, steam valves and generators, lighting and controls, water treatment, heat exchangers, insulation and roofing, unit heaters, fans, doors/windows/dock shelters, building retrofit/design, solar and wind technology, and cogeneration, etc.)
Miscellaneous: Publication for corporation and technical managers in business and industry.

156 **ENERGY POLICY**

Butterworth Scientific, Ltd.
P.O. Box 63, Westbury House
Bury Street
Guildford, Surrey GU2 5BH, England

Editor: Lyndon Driscoll
Phone: Guildford 0483-31261
ISSN: 0301-4215 **Circulation:** 7,000 **Frequency:** BM
Style Sheet: Yes
Revised Theses or Diss.: No **Manuscript Copies:** 3
Review Period: 1-3 months (PR) **Lagtime:** 9-12 months
Types of Articles: Research, review/overview papers, think pieces, rejoinders, letters, conference reports, and book reviews.
Major Contents: Politics, economics, planning and social aspects of energy; international energy policy; national studies; third world energy; structure of energy markets; conservation; comparative economics; energy/GDP interactions; energy planning under uncertainty; energy sector studies; environmental aspects; and forecasting. Economic analysis of all energy sources.

157 **ENERGY PROCESSING/CANADA**

205 9th Avenue, S.E.
Suite 503
Calgary, Alberta
T2G OR3, Canada

Editor: Scott Jeffrey
Phone: 403-265-4750
ISSN: 0319-5759 **Circulation:** 6,800 **Frequency:** BM
Style Sheet: Yes
Revised Theses or Diss.: Yes **Manuscript Copies:** 2

Review Period: 1 month **Lagtime:** 2 months
Types of Articles: Research, projects, and technical papers on downstream oil and gas industry.
Major Contents: Refineries; petrochemicals; new technology; natural gas processing; heavy oil; oil sands; pipelines; process control; alternate energy; northern exploration and development; instrumentation; energy conservation; and computers.
Miscellaneous: Each issue features specific aspects of the industry. The annual "Editorial Lineup" may be obtained from the editor. Papers must be submitted two months before the closing date of each issue.

158 **ENERGY PROGRESS**
Gas Processor Association
1812 First Place
Tulsa, OK 74103

Editor: Carl Sutton
Phone: 918-582-5112
ISSN: 0278-4521 **Circulation:** 3,900 **Frequency:** Q
Style Sheet: Yes
Revised Theses or Diss.: N/A **Manuscript Copies:** 3
Review Period: 3-6 months (PR) **Lagtime:** 6 months
Types of Articles: Research, projects, and general.
Major Contents: Materials, resources, and processes affecting energy.

159 **ENERGY RESEARCH REPORTS**
17 Langdon Avenue
Box 157
Watertown, MA 02172

Editor: D.B. Doucette
Phone: 617-872-8200
ISSN: 0190-4876 **Circulation:** N/A **Frequency:** SM
Style Sheet: Yes
Revised Theses or Diss.: No **Manuscript Copies:** 1
Review Period: 1-2 months **Lagtime:** N/A
Types of Articles: Projects, reviews, research, field tests, and project status.
Major Contents: All energy-related technologies.

160 **ENERGY REVIEW**
Greater Columbia Magazine Company
1120 Broad River Road
Columbia, SC 29210

Editor: Joao Pimentel, c/o Olade, Olade Building, P.O. Box 6413 C.C.I., Quito, Ecuador
Phone: 538-785
ISSN: N/A **Circulation:** 2,000 **Frequency:** 3/year
Style Sheet: No
Revised Theses or Diss.: No **Manuscript Copies:** 1
Review Period: 3-4 months **Lagtime:** 4-8 months
Types of Articles: Specialized articles on Latin American Countries energy policies, planning, research activities, and projects of particular interest.
Major Contents: New and renewable sources of energy; non-renewable sources of energy; energy planning and conservation; energy economics; and energy technology.

161 **ENERGY SOURCES**
Crane, Russak and Company, Inc.
3 E. 44th Street
New York, NY 10017

Editor: Dr. T.F. Yen, School of Engineering, University of Southern California (USC), University Park, Los Angeles, CA 90089-0242
Phone: 212-867-1490
ISSN: 0090-8212 **Circulation:** N/A **Frequency:** Q
Style Sheet: Yes
Revised Theses or Diss.: Yes **Manuscript Copies:** 3
Review Period: 3 months (PR) **Lagtime:** 6-9 months
Types of Articles: Research.
Major Contents: All sources of energy: international; and interdisciplinary; analytical studies of energy systems; problems from resource assessment to final consumption; and constraints on energy systems.
Miscellaneous: Interdisciplinary investigation of new trends and solutions to resolve the crisis of a worldwide energy shortage.

162 **ENERGY SYSTEMS AND POLICY**
Crane, Russak and Company, Inc.
3 E. 44th Street
New York, NY 10017

Editor: S. William Gouse, Jr., The Mitre Corporation, 1820 Dolley Madison Boulevard, McLean, VA 22102
Phone: 212-867-1490
ISSN: 0090-8347 **Circulation:** N/A **Frequency:** Q
Style Sheet: Yes

Revised Theses or Diss.: Yes **Manuscript Copies:** 3
Review Period: 3 months (PR) **Lagtime:** 6-9 months
Types of Articles: Research.
Major Contents: International in scope. Interdisciplinary science and technology for conventional and new sources of energy; development of these sources into practical fuels; environmental, legislative, and social concerns.

163 ENGINEERING AND MINING JOURNAL (E&MJ)

McGraw-Hill Publishing Company
1221 Avenue of the Americas
New York, NY 10020

Editor: Stanley H. Dayton
Phone: 212-512-2264
ISSN: 0095-8948 **Circulation:** 24,000 **Frequency:** M
Style Sheet: No
Revised Theses or Diss.: No **Manuscript Copies:** 2
Review Period: 2-3 months **Lagtime:** 2-3 months
Types of Articles: Articles and news.
Major Contents: Focus is on mining and metallurgy, including smelting and refining worldwide. Covers mining from exploration to final processing and marketing, as well as coal on an international scale and uranium.

164 ENGINEERING ECONOMIST, THE

Institute of Industrial Engineers
25 Technology Park/Atlanta
Norcross, GA 30092

Editor: Gerald Thuesen
Phone: 404-449-0460
ISSN: 0013-791X **Circulation:** 2,400 **Frequency:** Q
Style Sheet: Yes
Revised Theses or Diss.: N/A **Manuscript Copies:** 3
Review Period: 6 months (PR) **Lagtime:** 12 months
Types of Articles: Articles, reviews, case studies, and comments on articles and reviews.
Major Contents: Refereed journal devoted to the problems of capital investment and inventory as they relate to engineering and industrial development.

165 **ENGINEERING JOURNAL**

EIC Publishers, Inc.
700 EIC Building
2050 Mansfield Street
Montreal, Quebec H3A 1Y9, Canada

Editor: Robert J. Thibault, Corpus Information Services, Ltd., 1450 Don Mills Road, Don Mills, Ontario M3B 2X7, Canada
Phone: 416-445-7101
ISSN: 0013-8010 **Circulation:** 19,000 **Frequency:** BM
Style Sheet: No
Revised Theses or Diss.: Yes **Manuscript Copies:** N/A
Review Period: 1-2 months **Lagtime:** N/A
Types of Articles: Research, informative articles, and news.
Major Contents: Covers all aspects of engineering. Issues may be theme oriented.

166 **ENHANCED ENERGY RECOVERY NEWS**

Business Commmunication Company, Inc.
P.O. Box 2070 C
Stamford, CT 06906

Editor: Donald Saxman
Phone: 713-499-3932
ISSN: 0271-7085 **Circulation:** N/A **Frequency:** M
Style Sheet: No
Revised Theses or Diss.: No **Manuscript Copies:** N/A
Review Period: N/A **Lagtime:** 2 months
Types of Articles: News releases.
Major Contents: Enhanced energy.
Miscellaneous: A place to announce new enhanced recovery techniques but not to detail them.

167 **ENVIRONMENT**

H. Dwight Reid, Educational Foundation
4000 Albemarle Street, N.W.
Washington, DC 20016

Editor: Jane Scully, Managing Editor
Phone: 202-362-6445
ISSN: 0013-9157 **Circulation:** 20,000 **Frequency:** 10/year
Style Sheet: Yes
Revised Theses or Diss.: Yes **Manuscript Copies:** 3

Review Period: 2 months (PR) **Lagtime:** 3-4 months
Types of Articles: Research, projects, reviews, and public issues.
Major Contents: Interrelationship between science and technology and the physical, biological, and social environment.

168 **ENVIRONMENTAL ACTION**
Environmental Action, Inc.
1525 New Hampshire Avenue, N.W.
Washington, DC 20036

Editor: Rose Audette
ISSN: N/A **Circulation:** 25,000 **Frequency:** BM
Style Sheet: Yes
Revised Theses or Diss.: N/A **Manuscript Copies:** N/A
Review Period: 6 months **Lagtime:** 3-12 months
Types of Articles: News, short research reviews as it relates to larger problems, and book reviews.
Major Contents: A survey of environmental problems, issues, and solutions. Deals with toxic wastes, power, conservation, pollution and other matters.
Miscellaneous: Not scientific in approach. Laymen oriented. Incorporates items of local interest.

169 **ENVIRONMENTAL GEOLOGY AND WATER SCIENCES**
Springer-Verlag Publishers
175 Fifth Avenue
New York, NY 10010

Editor: Philip E. LaMoreaux, Editor-in-Chief, P.O. Box 2310, Tuscaloosa, Al 35403
Phone: 212-460-1500
ISSN: 0099-0094 **Circulation:** 778 **Frequency:** Q
Style Sheet: See Miscellaneous
Revised Theses or Diss.: Yes **Manuscript Copies:** 3
Review Period: N/A **Lagtime:** N/A
Types of Articles: Research, editorials, and book reviews.
Major Contents: Broad topics in earth sciences dealing with the interaction between humanity and the earth. Geologic hazards and geologic processes that affect people; management of geologic resources including land, water, air, and minerals including fuels; natural and man-made pollutants in the geologic environment; and environmental impact studies.
Miscellaneous: Suggestions to authors of the reports of the United States Geological Survey.

170 ENVIRONMENTAL SCIENCE AND TECHNOLOGY

American Chemical Society
1155 16th Street, N.W.
Washington, DC 20036

Editor: Dr. Russell F. Christman
Phone: 202-872-4600
ISSN: 0013-936X **Circulation:** 14,000 **Frequency:** M
Style Sheet: Yes
Revised Theses or Diss.: Yes **Manuscript Copies:** 3
Review Period: 3-5 months (PR) **Lagtime:** 2-3 months
Types of Articles: Projects, feature articles, critical reviews, current research, and correspondence.
Major Contents: Aspects of the environment and its control by scientific, engineering and political means. Original chemical research, engineering developments, and technico-economic studies in the fields of sciences directly related to man's environment. These studies primarily address the fundamental and applied aspects of water, air, and waste chemistry on the environment and its control by scientific, engineering and political means.
Miscellaneous: Address feature articles to Managing Editor; research papers to Manager, Manuscript Reviewing Office.

171 EPISODES

International Union of Geological Science
601 Booth Street, Room 177
Ottawa, Ontaria
K1A 0E8, Canada

Editor: Dr. A. R. Berger
Phone: 613-995-4927
ISSN: 0705-3797 **Circulation:** 3,000 **Frequency:** Q
Style Sheet: N/A
Revised Theses or Diss.: N/A **Manuscript Copies:** N/A
Review Period: N/A (PR) **Lagtime:** N/A
Types of Articles: Research, book reviews, conference reports, and news reports.
Major Contents: Developments of regional and global importance in the earth sciences.
Miscellaneous: Formerly: INTERNATIONAL UNION OF GEOLOGICAL SCIENCES. GEOLOGICAL NEWSLETTER, ISSN: 0047-1267.

172 **EUROPEAN POWER NEWS**

Industrial Newspapers, Ltd.
2 Queensway House
Redhill, Surrey RH1 1QS
England

Editor: C. McCombe
Phone: 0737 68611
ISSN: 0261-8214 **Circulation:** 11,100 **Frequency:** 8/year
Style Sheet: Yes
Revised Theses or Diss.: No **Manuscript Copies:** 1
Review Period: 1 month **Lagtime:** 3 months
Types of Articles: Projects and reviews.
Major Contents: Independent power generation products throughout western Europe; electrotechnical standards; advanced control systems for generating sets; marinization and tropicalization of generating sets.
Miscellaneous: Articles concentrate on European power developments. Research developments are presented in semi-technical terms. Credits are given for articles submitted. Articles are commissioned approximately three months before publications.

173 **EXPERIENTIA: MONTHLY JOURNAL FOR NATURAL SCIENCES**

Birkhauser Verlag
P.O. Box 133
CH-4010 Basel
Switzerland

Editor: H. Mislin and H.P. Von Hahn
Phone: 061-73-53-00
ISSN: 0014-4754 **Circulation:** 2,500 **Frequency:** M
Style Sheet: Yes
Revised Theses or Diss.: No **Manuscript Copies:** 3
Review Period: 3-4 months (PR) **Lagtime:** 3-4 months
Types of Articles: Reviews, short communications, mini-reviews, and multi-author reviews (solicited).
Major Contents: Life sciences with emphasis on interdisciplinary aspects.

174 **FATIGUE AND FRACTURE OF ENGINEERING MATERIALS AND STRUCTURES**

Pergamon Press
Maxwell House, Fairview Park
Elmsford, NY 10523

Editor: Professor K.J. Miller, Department of Mechanical Engineering, University of Sheffield, Mappin Street, Sheffield S1 3JD, United Kingdom
Phone: 914-592-7700
ISSN: 0160-4112 **Circulation:** 2,000 **Frequency:** Q
Style Sheet: Yes
Revised Theses or Diss.: Yes **Manuscript Copies:** 2
Review Period: 1-2 months (PR) **Lagtime:** 2-6 months
Types of Articles: Research, articles, and reviews.
Major Contents: Current knowledge on fatigue and fracture that will assist engineers or scientists in all disciplines.

175 FERROELECTRICS AND RELATED MATERIALS

Gordon and Breach Science Publishers, Inc.
1 Park Avenue
New York, NY 10010

Editor: G.W. Taylor
Phone: 609-890-7915
ISSN: 0015-0193 **Circulation:** N/A **Frequency:** 7/year
Style Sheet: American Institute of Physics.
Revised Theses or Diss.: N/A **Manuscript Copies:** 3
Review Period: 1 month (PR) **Lagtime:** 6 months
Types of Articles: Research and reviews.
Major Contents: Ferroelectrics, ferroelastics, polar dielectrics, electro-optics, piezoelectrics, nonlinear dielectrics, liquid crystals, and pyroelectrics.

176 FIRST BREAK

Blackwell Scientific Publications, Ltd.
Osney Mead, Oxford OX2 0EL
England

Editor: Neil Goulty, Department of Geological Sciences, University of Durham, South Road, Durham DH1 3LE, United Kingdom
Phone: 0385 64971
ISSN: 0263-5046 **Circulation:** 4,000 **Frequency:** M
Style Sheet: Yes
Revised Theses or Diss.: No **Manuscript Copies:** 4
Review Period: 1 month (PR) **Lagtime:** 4 months
Types of Articles: Original research, reviews, case histories, and state-of-the-art.
Major Contents: A wide range of articles are accepted provided they are written for intelligent, numerate but non-mathematical geoscientists

and geophysicists. Geophysics, theory practice, instrumentation, interpretation, particularly exploration geophysics and any related areas, e.g. geology, computing etc. are of primary interest.

Miscellaneous: There are no preferred topics; they are either selected from whatever is submitted or commissioned.

177 **FLUID PHASE EQUILIBRIA**

Ecole des Mines
60 Boulevard, Saint-Michel
75006 Paris, France

Editor: Professor H. Renon
ISSN: N/A **Circulation:** N/A **Frequency:** M
Style Sheet: No
Revised Theses or Diss.: No **Manuscript Copies:** 3
Review Period: 2-3 months (PR) **Lagtime:** 6-8 months

Types of Articles: Research, reviews, short communications, and letters to the editor.

Major Contents: Multidisciplinary journal for experimental, theoretical and applied aspects of fluid phase equilibria. Topics include pure and mixed materials, thermodynamic properties, chemical thermodynamics, applications in engineering e.g., separation processes, thermal engines, and oil recovery.

Miscellaneous: This journal provides 50 free reprints free of charge.

178 **FOOD TECHNOLOGY**

Institute of Food Technologists
221 N. LaSalle Street
Chicago, IL 60601

Editor: John B. Klis
Phone: 312-782-8424
ISSN: 0015-6639 **Circulation:** 20,000 **Frequency:** M
Style Sheet: No
Revised Theses or Diss.: No **Manuscript Copies:** N/A
Review Period: 2-3 months **Lagtime:** N/A

Types of Articles: Book reviews, news and feature articles, patents, professional placement information, and professional directory.

Major Contents: The development of new and improved food sources, products, and processes, their proper utilization by industry and the consumer, and their effective regulation by government agencies. Topics include, but are not limited to: biochemistry, biology, consumerism, education, energy analysis, engineering, equipment, experimental design, food safety, food service, history, information retrieval and documentation, laboratory equipment and instrumentation, management, marketing, microbiology, nutrition, packaging, pollution control, processing, process instrumentation and control, product development, quality assurance, regulatory matters, and research.

Miscellaneous: An outline of proposed article is requested prior to submission.

179 **FUEL OIL NEWS**

Publex Corporation
P.O. Box 360
Whitehouse, NJ 08888

Editor: George Schultz
Phone: 201-534-4156
ISSN: 0378-3820 **Circulation:** 17,000 **Frequency:** M
Style Sheet: No
Revised Theses or Diss.: No **Manuscript Copies:** 1
Review Period: 2-3 months **Lagtime:** 2 months
Types of Articles: Current awareness.
Major Contents: Fuel oil delivery; heating equipment.
Miscellaneous: Directed to fuel oil dealers.

180 **FUEL PROCESSING TECHNOLOGY**

Elsevier Science Publishers B.V.
P.O. Box 330
1000 AH Amsterdam
The Netherlands

Editor: L.L. Anderson, Department of Fuels Engineering, Browning Mineral Science Building, University of Utah, Salt Lake City, UT 84112
ISSN: 0378-3820 **Circulation:** N/A **Frequency:** BM
Style Sheet: Yes
Revised Theses or Diss.: Yes **Manuscript Copies:** 3
Review Period: 1-2 months (PR) **Lagtime:** 2-3 months
Types of Articles: Primary research, scholarly reviews, short communications, and letters to the editor.
Major Contents: Scientific and technical aspects of fuel processing technology including: fuels to other fuels; chemicals and by-products; coal, oil, tar, sands, and peat, including optimization, catalysis, feeding, product separation, catalyst recovery, economics, pollution control, liquefaction, gasification, solvent extraction, desulfurization and other upgrading processes.

181

FUEL: THE SCIENCE AND TECHNOLOGY OF FUEL AND ENERGY

Butterworth Scientific, Ltd.
Box 63, Westbury House
Bury Street
Guildford, Surrey GU2 5BH, England

Editor: Herb Retcofsky, Butterworth Scientific, Ltd., US Department of Energy, P.O. Box 10940, Pittsburgh, PA 15236
Phone: 412-675-5786
ISSN: 0016-2361 **Circulation:** N/A **Frequency:** M
Style Sheet: Yes
Revised Theses or Diss.: Yes **Manuscript Copies:** 3
Review Period: 2-3 months **Lagtime:** 6-8 months
Types of Articles: Research, short communications, and correspondence.
Major Contents: All aspects of the science and technology of fuel and energy: alternative fuels, processes, analysis and properties of basic fuels/materials; coal chemistry, oil shales, tar sands, new uses of coal/new sources of fuel.
Miscellaneous: Jayne Holder, Executive Editor, FUEL, Butterworth Scientific, Ltd., P.O. Box 63, Westbury House, Bury Street, Guildford, Surrey GU2 5BH, England. Phone: 0483-31262.

182

FUSION

Fusion Energy Foundation
Box 1438, Radio City Street
New York, NY 10101

Editor: Marjorie Hecht, Managing Editor
Phone: 212-247-8439
ISSN: 0148-0537 **Circulation:** 150,000 **Frequency:** BM
Style Sheet: Yes
Revised Theses or Diss.: Yes **Manuscript Copies:** 1
Review Period: 2 months (PR) **Lagtime:** 3-4 months
Types of Articles: Research, projects, reviews on developments, innovations and applications, and interviews.
Major Contents: Fusion and nuclear energy; developments "at the frontiers of science and energy" including astrophysics, biology, and medicine; innovations in technology; industrial and agricultural applications of technology; children's section "The Young Scientist"; the history of science; beam defense developments; economic development; space science; science and technology development internationally.

Miscellaneous: FUSION is published also in German, French, Italian, Spanish, and Swedish, quarterly and in English as FUSION ASIA, (published in New Delhi, India) quarterly.

183 **FUSION TECHNOLOGY**
American Nuclear Society
555 N. Kensington Avenue
La Grange Park, IL 60525

Editor: George H. Miley, University of Illinois, 214 Nuclear Engineering Laboratory, 103 South Goodwin Avenue, Urbana, IL 61801
Phone: 217-333-3772
ISSN: 0272-3921 **Circulation:** 2,000 **Frequency:** BM
Style Sheet: Yes
Revised Theses or Diss.: Yes **Manuscript Copies:** 3
Review Period: 3 months (PR) **Lagtime:** 5-6 months
Types of Articles: Research, projects, reviews, technical notes, and letters to the editor.
Major Contents: All phases of applications of fundamental research to fusion technology.
Miscellaneous: Published under the joint ownership of the American Nuclear Society and the European Nuclear Society. Formerly: NUCLEAR TECHNOLOGY/FUSION.

184 **GAS DIGEST**
Tri-Plek Productions
Box 35819
Houston, TX 77235

Editor: Ken Kridner
Phone: 713-723-7456
ISSN: 0161-4851 **Circulation:** 9,200 **Frequency:** M
Style Sheet: No
Revised Theses or Diss.: Yes **Manuscript Copies:** 1
Review Period: 2 weeks **Lagtime:** 1-4 months
Types of Articles: Research, articles, reviews equipment, news and uses, and innovative processes.
Major Contents: Topics of interest to people who work in the national gas area. Primarily operations oriented.
Miscellaneous: All articles are semi-technical in presentation.

185 **GAS ENERGY REVIEW**
American Gas Association
1515 Wilson Boulevard
Arlington, VA 22209

Editor: Jeffrey L. Wingenroth
Phone: 703-841-8506
ISSN: 8756-5471 **Circulation:** 2,000 **Frequency:** M
Style Sheet: No
Revised Theses or Diss.: No **Manuscript Copies:** 2
Review Period: 1 month **Lagtime:** 2-3 months
Types of Articles: Analyses of gas industry issues performed by the policy evaluation and analysis group of the American Gas Association. On occasion, articles authored by other divisions of AGA or by officials of gas industry companies are included.
Major Contents: Analyses of natural gas supply, demand, prices, financial issues, legislative and regulatory developments, new technology (both supply oriented and utilization oriented), and key policy issues.
Miscellaneous: Outside manuscripts are accepted but must have prior authorization.

186 **GEOCHIMICA ET COSMOCHIMICA ACTA**

Pergamon Press
Maxwell House, Fairview Park
Elmsford, NY 10523

Editor: Professor Denis M. Shaw, Department of Geology, McMaster University, 1280 Main Street, Hamilton, Ontario, L85 4M1, Canada
Phone: 416-525-9140
ISSN: 0016-7037 **Circulation:** 3,800 **Frequency:** M
Style Sheet: Inside back cover of journal.
Revised Theses or Diss.: Yes **Manuscript Copies:** 5
Review Period: 4 months (PR) **Lagtime:** 1-6 months
Types of Articles: Original research, articles, notes, book reviews, and letters to the editor.
Major Contents: Geochemistry and cosmochemistry such as meteorites, geochronology, marine geochemistry, organic geochemistry, general geochemistry, and comments on published papers.
Miscellaneous: An official journal of the Geochemical Society and the Meteoritical Society. Manuscripts may be submitted in English, French or German.

187 **GEOEXPLORATION**

Elsevier Science Publishers B.V.
P.O. Box 211
1000 AE Amsterdam
The Netherlands

Editor: Dr. D.S. Parasnis, Editor-in-Chief, Department of Applied Geophysics, University of Lulea, 95187 Lulea, Sweden
ISSN: 0016-7142 **Circulation:** N/A **Frequency:** Q
Style Sheet: Yes
Revised Theses or Diss.: Yes **Manuscript Copies:** 3
Review Period: 2-3 months (PR) **Lagtime:** N/A
Types of Articles: Research and reviews.
Major Contents: Mining and technical geophysics and related subjects.
Miscellaneous: The official language is English but French and German articles are occasionally accepted. Any non-English language article should be prefaced by an abstract in English. U.S. and Canadian inquiries may be addressed to Elsevier/North-Holland Inc., Journal Information Center, 52 Vanderbilt Avenue, New York, NY 10017, Phone: 212-867-9040.

188 **GEOLOGICAL JOURNAL**
John Wiley and Sons, Ltd.
Baffins Lane, Chichester
Sussex P01 IUD
England

Editor: Dr. P.J. Brenchley, Geological Journal, Department of Geology, University of Liverpool, Liverpool L693BX, England
Phone: 051-709-6022
ISSN: 0072-1050 **Circulation:** 1,000 **Frequency:** Q
Style Sheet: Yes
Revised Theses or Diss.: No **Manuscript Copies:** 3
Review Period: 1-2 months (PR) **Lagtime:** 6-9 months
Types of Articles: Original research and reviews.
Major Contents: All branches of the geological sciences particularly regional case studies and research on sedimentology, palaeontology, structural geology, geophysics, geochemistry, metamorphic geology, and igneous petrology.
Miscellaneous: U.S. information regarding this publication may be obtained by writing Expediters of the Printed Word, Ltd., 515 Madison Avenue, New York, NY 10002.

189 **GEOLOGICAL MAGAZINE**
Cambridge University Press
Edinburgh Building
Shaftesbury Road
Cambridge CB2 2RU, England

Editor: W.B. Harland, Geological Magazine, Department of Earth Sciences, Downing Street, Cambridge CB2 3EQ, England
Phone: 0223-355463
ISSN: 0016-7568 **Circulation:** N/A **Frequency:** BM
Style Sheet: Yes
Revised Theses or Diss.: No **Manuscript Copies:** 3
Review Period: 1 month (PR) **Lagtime:** 4-12 months
Types of Articles: Research, projects, and reviews.
Major Contents: All areas of geology.

190 GEOLOGICAL SOCIETY OF INDIA JOURNAL

Geological Society of India
B.B.D. Press
Bangalore, 560 053
South India

Editor: B.P. Radhakrishna
ISSN: 0016-7622 **Circulation:** 1,300 **Frequency:** M
Style Sheet: Yes
Revised Theses or Diss.: Yes **Manuscript Copies:** 2
Review Period: 6 months **Lagtime:** 12 months
Types of Articles: Reviews, research, short communications, and book reviews.
Major Contents: Crustal evolution; metallogeny; petrology; Indian geology; paleontology and stratigraphy; and geomorphology.

191 GEOLOGY

Geological Society of America, Inc.
P.O. Box 9140
3300 Penrose Place
Boulder, CO 80301

Editor: Eldridge M. Moores
Phone: 303-447-2020
ISSN: 0091-7613 **Circulation:** 12,500 **Frequency:** M
Style Sheet: Yes
Revised Theses or Diss.: N/A **Manuscript Copies:** 3
Review Period: 1-3 months (PR) **Lagtime:** 3-4 months
Types of Articles: Research and reviews of general interest for geologists.
Major Contents: All aspects of geology.

192 GEOPHYSICAL AND ASTROPHYSICAL FLUID DYNAMICS

Gordon and Breach Science Publishers, Ltd.
1 Bedford Street
London WC2E 9TP
England

Editor: P.H. Roberts, School of Mathematics, The University of New Castle upon Tyne, New Castle upon Tyne NE1 7RU, England
Phone: 01-836-5125
ISSN: 0309-1929 **Circulation:** N/A **Frequency:** 4/year
Style Sheet: Yes
Revised Theses or Diss.: N/A **Manuscript Copies:** 3
Review Period: N/A (PR) **Lagtime:** N/A
Types of Articles: Original research, observational studies, short communications, survey articles, conference reports, and book reviews.
Major Contents: Fluid mechanics of the earth and planets, including oceans, atmospheres and interiors, and the fluid mechanics of the sun, stars and other astrophysical objects. Magnetohydrodynamic behaviors are investigated and experimental, theoretical, and numerical studies of rotating stratified and convecting fluids of general interest to geophysicists and astrophysicists appear.
Miscellaneous: Promotes exchange of ideas between researchers and specialists within the fields of meteorology, oceanography, planetary physics, applied math, and fluid dynamics. North American contributions may be sent to: F.H. Busse, Institute of Geophysics and Planetary Physics, University of California, Los Angeles, CA 90024.

193 GEOPHYSICAL PROSPECTING

Blackwell Scientific Publications, Ltd.
Osney Mead Oxford
OX2 0EL, England

Editor: Dr. K. Helbig, Geophysical Prospecting, Editorial Office, Couwenhoven 62-49, 3703 HN ZEIST, The Netherlands
Phone: 034-045-1019
ISSN: 0016-8025 **Circulation:** 4,300 **Frequency:** 8/year
Style Sheet: Yes
Revised Theses or Diss.: Yes **Manuscript Copies:** 3
Review Period: 2-3 months (PR) **Lagtime:** 6 months
Types of Articles: Research, projects, reviews, and news articles.
Major Contents: Geophysics relating to exploration and related sciences.
Miscellaneous: An official journal of the European Association of Exploration Geophysicists.

194 **GEOPHYSICAL RESEARCH LETTERS**

American Geophysical Union
2000 Florida Avenue, N.W.
Washington, DC 20009

Editor: Dr. Douglas Torr
Phone: 202-462-6903
ISSN: 0094-8276 **Circulation:** 4,500 **Frequency:** M
Style Sheet: N/A
Revised Theses or Diss.: N/A **Manuscript Copies:** 4
Review Period: 1 week (PR) **Lagtime:** 2-3 months
Types of Articles: Research.
Major Contents: Broad geophysical interest.
Miscellaneous: Send three manuscript copies to the co-editor, and one to American Geophysical Union headquarters.

195 **GEOPHYSICAL SURVEYS**

D. Reidel Publishing Company
Journals Department
P.O. Box 17
3300 AA Dordrecht, Holland

Editor: Professor K. Creer, Department of Geophysics, University of Edinburgh, Mayfield Road, James Clerk Maxwell Building, Edinburgh EH9 2J2, United Kingdom
Phone: 031-667-1081-2952
ISSN: 0046-5763 **Circulation:** N/A **Frequency:** Q
Style Sheet: Yes
Revised Theses or Diss.: No **Manuscript Copies:** 3
Review Period: 1 month **Lagtime:** 1-6 months
Types of Articles: Expository review articles. No original research.
Major Contents: Coverage is both academic and applied geophysical research in its broadest sense. Subjects within the scope of the publication are geodesy, seismology, meteorology, geomagnetism, paleomagnetism oceanography, petrology, hydrology, tectonics, the moon, sun and the planets, and new applications of basic physical principles to applied geophysics.
Miscellaneous: Publishes only invited survey papers by researchers with an established international reputation. These articles are written for non-specialists and should contain ample explanatory material.

196 **GEOPHYSICAL TRANSACTIONS**

Eotvos Lorand Geophysics
Institute of Hungary
P.O.B. 35
Budapest, H-1440, Hungary

Editor: Dr. Eva Kilenyi
Phone: 36-1-632-835
ISSN: 0016-7177 **Circulation:** 1,400 **Frequency:** Q
Style Sheet: Yes
Revised Theses or Diss.: Yes **Manuscript Copies:** 2
Review Period: 2 months (PR) **Lagtime:** 6 months
Types of Articles: Research, case histories, monographs, and occasional book reviews.
Major Contents: Regional geology; tectonics; paleomagnetism; geothermics; seismology; wave theory; applied geophysics; gravimetry; magnetic methods; geoelectrics; seismics; mining geophysics; engineering geophysics; in-mine geophysics; and well logging.
Miscellaneous: This journal is the same as GEOFTZIKAI KOZLEMENYEK.

197 GEOPHYSICS
Society of Exploration Geophysicists
8801 S. Yale
Tulsa, OK 74137

Editor: Stanley H. Ward, University of Utah Research Institute, Earth Science Laboratory, 391 Chipeta Way, Suite C, Salt Lake City, Utah 84108
Phone: 801-524-3454
ISSN: 0016-8033 **Circulation:** 22,000 **Frequency:** M
Style Sheet: See Miscellaneous
Revised Theses or Diss.: Yes **Manuscript Copies:** 5
Review Period: 7-11 months (PR) **Lagtime:** 5 months
Types of Articles: Research, tutorial papers, reviews, case histories of exploration, integrated studies, model studies, applied instrumentation, computer programs, and reports of research workshops.
Major Contents: Seismic studies (land marine); vertical seismic profiling; mining; electrical and electromagnetic; remote sensing; well logging; gravity and magnetic surveying; marine studies (sea floor, waste storage).
Miscellaneous: Chicago Manual of Style, "SEG Instructions to Authors" and SED Encyclopedia Dictionary of Exploration Geophysics.

198 GEOPHYSICS: THE LEADING EDGE OF EXPLORATION
Society of Exploration Geophysicists
8801 S. Yale
Tulsa, OK 74137

Editor: Theodore Barrington, P.O. Box 3098, Tulsa, OK 74101
Phone: 918-493-3516
ISSN: 0732-989X **Circulation:** 20,205 **Frequency:** M
Style Sheet: See Miscellaneous
Revised Theses or Diss.: Yes **Manuscript Copies:** 1
Review Period: 1 month (PR) **Lagtime:** 2 months
Types of Articles: Research, projects, reviews, and profiles of well-known earth scientists.
Major Contents: Exploration geophysics (nonmathematical), other earth sciences, general energy, and biographical profiles.
Miscellaneous: Chicago Manual of Style, "SEG Instructions to Authors" and SED Encyclopedia Dictionary of Exploration Geophysics.

199 **GEOPOLITICS OF ENERGY**
Conant and Associates, Ltd.
9901 Phoenix Lane
Great Falls, VA 22066

Editor: Melvin A. Conant
Phone: 202-783-7456
ISSN: N/A **Circulation:** N/A **Frequency:** N/A
Style Sheet: N/A
Revised Theses or Diss.: N/A **Manuscript Copies:** N/A
Review Period: N/A **Lagtime:** N/A
Types of Articles: Research and reviews.
Major Contents: Covers information and opinion on the political and economic factors affecting energy and their impact on national energy policies and the international environment. Reports on developments in producer and consumer countries but also provides critiques and analyses of current events as they affect those countries now and in the future.
Miscellaneous: This journal is read worldwide by business and financial executives as well as government policymakers. The Editorial Committee includes experts on energy, government, and industry, with experience in the Middle East, Far East, Europe, and Africa.

200 **GEOTHERMAL ENERGY**
Geothermal World Corporation
5762 Firebird Court, Mission Oaks
Camarillo, CA 93010

Editor: Karen Lee Darwin
Phone: 805-482-6288

ISSN: 0093-5700 **Circulation:** 6,000 **Frequency:** M
Style Sheet: Yes
Revised Theses or Diss.: Yes **Manuscript Copies:** 2
Review Period: 1-3 months **Lagtime:** 1-2 months
Types of Articles: Research, projects, reviews, and technology.
Major Contents: All aspects geothermal energy and volcanoes.

201 GEOTHERMICS: INTERNATIONAL JOURNAL OF GEOTHERMAL RESEARCH AND ITS APPLICATIONS
Pergamon Press, Maxwell House
Fairview Park
Elmsford, NY 10523

Editor: E. Barbier, Geothermics-Editorial Office, c/o Instituto Internazionale per le, Ricerche Geotermiche, Via del Buongusto 1, 56100 Pisa, Italy
Phone: 050-41503/46069
ISSN: 0375-6505 **Circulation:** 1,075 **Frequency:** Q
Style Sheet: Yes
Revised Theses or Diss.: Yes **Manuscript Copies:** 2
Review Period: 1-3 months **Lagtime:** 8-10 months
Types of Articles: Research, reviews, and news in the geothermal energy sector
Major Contents: Exploration and exploitation of geothermal energy. All aspects of geothermal energy research and development.

202 GROUNDSWELL
Nuclear Information and Resource Service
1536 16th Street, N.W.
Washington, DC 20036

Editor: The Nuclear Information and Resource Service, 1346 Conneticut Avenue, N.W., 4th Floor, Washington, DC 20036
Phone: 202-296-7552
ISSN: 0162-7899 **Circulation:** 1,000 **Frequency:** BM
Style Sheet: Yes
Revised Theses or Diss.: No **Manuscript Copies:** 1
Review Period: 1 month **Lagtime:** 2 months
Types of Articles: Research and analysis.
Major Contents: Nuclear power, radiation, and health.

203 GROWTH AND CHANGE
University of Kentucky
644 Maxwelton Court
Lexington, KY 40506

Editor: Charles Hultman
Phone: 606-257-3827
ISSN: 0017-4815 **Circulation:** 1,350 **Frequency:** Q
Style Sheet: Yes
Revised Theses or Diss.: No **Manuscript Copies:** 4
Review Period: 4-6 months (PR) **Lagtime:** 6 months
Types of Articles: Research, theory, and occasional reviews.
Major Contents: Public, urban, and regional policy, from the standpoint of economists, geographers, planners, and sociologists; location theory; migration studies; and regional and urban development.

204 **HEALTH PHYSICS**
Pergamon Press
Maxwell House, Fairview Park
Elmsford, NY 10523

Editor: Dr. Genevieve S. Roessler, Editor-in-Chief, Beth Magura, Managing Editor, 202 Nuclear Sciences Center, University of Florida, Gainesville, FL 32611
Phone: 904-392-1404
ISSN: 0017-9078 **Circulation:** 6,800 **Frequency:** M
Style Sheet: Yes
Revised Theses or Diss.: Yes **Manuscript Copies:** 3
Review Period: 1-3 months (PR) **Lagtime:** 7-8 months
Types of Articles: Original research and review papers on theoretical and applied aspects of radiation protection.
Major Contents: Theoretical and applied fields of health physics. Operation health physics (i.e. reactor H.P.); environmental monitoring; personnel monitoring; dosimetry; waste management; medical health physics; atomic bomb survivor studies; microwaves; ultrasound; lasers; radiobiology radiation bioeffects; plutonium handling; accident analysis; and fallout from weapons tests.

205 **HEAT TRANSFER ENGINEERING AN INTERNATIONAL QUARTERLY**
Hemisphere Publishing Corporation
1010 Vermont Avenue, N.W.
Washington, DC 20005

Editor: Kenneth J. Bell, Editor-in-Chief, School of Chemical Engineering, Oklahoma State University, Stillwater, OK 74074
Phone: 202-783-3985

ISSN: 0145-7632 **Circulation:** 2,010 **Frequency:** Q
Style Sheet: Yes
Revised Theses or Diss.: N/A **Manuscript Copies:** 3
Review Period: N/A (PR) **Lagtime:** N/A
Types of Articles: Refereed papers of original work, state-of-the-art reviews, reviews of fundamentals, news items, book reviews, announcements of meetings and educational programs, and advertising.
Major Contents: Topics include heat transfer practice, new developments in equipment or practices.
Miscellaneous: Aimed at the practicing engineer and others in the field of heat transfer.

206 **HEATING/PIPING/AIR CONDITIONING**
Penton-IPC, Reinhold Publishing Division
2 Illinois Center, Suite 1300
Chicago, IL 60601

Editor: Robert T. Korte
Phone: 203-348-7531
ISSN: 0017-940X **Circulation:** 41,000 **Frequency:** M
Style Sheet: No
Revised Theses or Diss.: Yes **Manuscript Copies:** 1
Review Period: 1-2 months **Lagtime:** 3-12 months
Types of Articles: Case studies.
Major Contents: Energy savings in HVAC design, energy economics; design and cost; technical, but practical systems approaches; innovation; and future trends.

207 **HELICTITE**
Speleological Research Council, Ltd.
P.O. Box 183
Broadway, N.S.W. 2007
Australia

Editor: J.M. James and H.J. Dyson
ISSN: 0017-9973 **Circulation:** 500 **Frequency:** SA
Style Sheet: Yes
Revised Theses or Diss.: Yes **Manuscript Copies:** 1
Review Period: 2 months **Lagtime:** 1-6 months
Types of Articles: Research, projects, write-ups, and some reviews of new books, etc. published occasionally.
Major Contents: Biospeleology; mineralogy; harst hydrology; speleochemistry; geomorphology; history speleological techniques. Preferred topics are earth sciences; meteorology; conservation; history; and car-

tography.

208 **HIGH TECHNOLOGY**

Technology Publishing Company
38 Commercial Wharf
Boston, MA 02110

Editor: Robert Haavind
Phone: 617-227-4700
ISSN: 0277-2981 **Circulation:** 300,000 **Frequency:** BM
Style Sheet: Yes
Revised Theses or Diss.: No **Manuscript Copies:** 1
Review Period: 1 month **Lagtime:** 3-5 months
Types of Articles: Research, projects, reviews of technical and practical nature.
Major Contents: Current developments in all types of technologies: practical information; developments in business, industry, military, aerospace, energy, and others.

209 **HIGH TEMPERATURE SCIENCE**

The Humana Press
P.O. Box 2148
Clifton, NJ 07015

Editor: John Margrave
Phone: 201-773-4389
ISSN: 0018-1536 **Circulation:** N/A **Frequency:** BM
Style Sheet: Yes
Revised Theses or Diss.: N/A **Manuscript Copies:** 3
Review Period: 2-4 weeks (PR) **Lagtime:** 3-6 months
Types of Articles: Research, reviews, and communications on new results or viewpoints.
Major Contents: Stresses theoretical and experimental work on high temperature phenomena, techniques, and materials. Examples include: thermodynamic and structural studies of high temperature systems by mass spectrometric, optical, x-ray, or electron defraction; electrochemical techniques; flame and explosion phenomena; electric arc chemistry and physics; and laser spectroscopy techniques.

210 **HYDROCARBON PROCESSING**

Gulf Publishing Company
P.O. Box 2608
Houston, TX 77001

Editor: Harold L. Hoffman
Phone: 713-529-4301
ISSN: 0018-8190 **Circulation:** 36,735 **Frequency:** M
Style Sheet: Yes
Revised Theses or Diss.: Yes **Manuscript Copies:** 1
Review Period: 1-2 months (PR) **Lagtime:** 2-12 months
Types of Articles: Articles of industrial significance concerning conversion of hydrocarbon, raw materials to fuels and other useful products; news; management techniques; and equipment descriptions.
Major Contents: Crude oil and natural gas processing; coal; tar sands; shale oil as feedstock; fueling; asphalt; wax; petrochemicals as products; engineering design; operations; and management economics.
Miscellaneous: To interest plant managers, chief engineers, project and process engineers, maintenance construction superintendents, chemical engineers, and purchasing personnel. Formerly: PETROLEUM REFINER.

211 **HYDROMETALLURGY**
Elsevier Science Publishers B.V.
P.O. Box 211
1000 AE Amsterdam
The Netherlands

Editor: G. M. Ritcey, Ottawa, Ontario, Canada
Phone: 020-5803-911
ISSN: 0340-386X **Circulation:** N/A **Frequency:** Q
Style Sheet: Yes
Revised Theses or Diss.: No **Manuscript Copies:** 3
Review Period: 2-4 weeks **Lagtime:** 4-6 months
Types of Articles: Research.
Major Contents: Aqueous process of metals with discussions on novel processes, process design, chemistry, modeling, control, economics, and interfaces between unit operations.

212 **IEE PROCEEDINGS PART A: MICROWAVES, OPTICS AND ANTENNAS**
Institution of Electrical Engineers
P.O. Box 26
Hitchin, Herts.
SG5 1SA, England

Editor: B. Dunkley
ISSN: N/A **Circulation:** 2,600 **Frequency:** BM
Style Sheet: Yes

Revised Theses or Diss.: Yes **Manuscript Copies:** 4
Review Period: 6 months (PR) **Lagtime:** 6 months
Types of Articles: Research.
Major Contents: Topics include: microwave techniques, components and instrumentation; microwave sources; integrated optics, arrays; aperture antennas; travelling-wave antennas; microwave network theory; and optical-fibre communications.

213 IEE PROCEEDINGS PART A: PHYSICAL SCIENCE, MEASUREMENT AND INSTRUMENTATION, MANAGEMENT AND EDUCATION, REVIEWS

Institution of Electrical Engineers
P.O. Box 26
Hitchin, Herts. SG5 1SA, England

Editor: B. Dunkley
ISSN: 0143-702X **Circulation:** 3,600 **Frequency:** BM
Style Sheet: Yes
Revised Theses or Diss.: Yes **Manuscript Copies:** 4
Review Period: 6 months (PR) **Lagtime:** 6 months
Types of Articles: Research, projects, and reviews.
Major Contents: Broadly based in-depth articles concerning science, measurement, management, design, medical and biological systems, education, and history. Examples include: materials, electric discharges, magnetism, measurement of electrical quantities, project management, engineering design concepts, biomedical engineering systems applications, curriculum development, and the history of electrical engineering.

214 IEE PROCEEDINGS PART B: ELECTRIC POWER APPLICATIONS

Institution of Electrical Engineers
P.O. Box 26
Hitchin, Herts. SG5 1SA
England

Editor: B. Dunkley
ISSN: 0143-7038 **Circulation:** 4,500 **Frequency:** BM
Style Sheet: Yes
Revised Theses or Diss.: Yes **Manuscript Copies:** 4
Review Period: 6 months (PR) **Lagtime:** 6 months
Types of Articles: Research and applications.
Major Contents: Topics concerning: electrical equipment design, application and development, rotating electrical machines; linear motors;

railway traction and signalling; rectifier and inverter-fed drives; applications of power-electronics devices; industrial and nonindustrial applications of electric power.

215 **IEE PROCEEDINGS PART C: GENERATION, TRANSMISSION AND DISTRIBUTION**

Institution of Electrical Engineers
Savoy Place
London WC2R 0BL
England

Editor: Bernard Dunkley, Managing Editor
Phone: 0438-313311
ISSN: 0143-7046 **Circulation:** 4,500 **Frequency:** BM*
Style Sheet: Yes
Revised Theses or Diss.: Yes **Manuscript Copies:** 4
Review Period: 3 months (PR) **Lagtime:** 6 months
Types of Articles: Research, projects, reviews, correspondence, and applications.
Major Contents: Generators; generation control; power systems design and analysis; protection plant design; switchgear; cables; transmission lines; equipment design and application; control and protection; as well as planning, operation, and management of power systems.
Miscellaneous: *The IEE Proceedings is published in nine parts. Six issues of each Part are published bimonthly each year.

216 **IEE PROCEEDINGS PART D: CONTROL THEORY AND APPLICATIONS**

Institution of Electrical Engineers
P.O. Box 26
Hitchin, Herts. SG5 1SA
England

Editor: B. Dunkley
ISSN: 0143-7054 **Circulation:** 3,000 **Frequency:** BM
Revised Theses or Diss.: Yes **Manuscript Copies:** 4
Review Period: 6 months (PR) **Lagtime:** 6 months
Types of Articles: Research.
Major Contents: Topics covered include: control systems; system modelling, design and implementation; identification; simulation; technological systems; economic systems; physiological systems; man-machine interfaces; robotics; and process control.

217 IEE PROCEEDINGS PART H: MICROWAVES, OPTICS AND ANTENNAS

Institution of Electrical Engineers
P.O. Box 26
Hitchin, Herts. SG5 1SA
England

Editor: B. Dunkley
ISSN: N/A **Circulation:** 2,600 **Frequency:** BM
Revised Theses or Diss.: N/A **Manuscript Copies:** 4
Review Period: 6 months (PR) **Lagtime:** 6 months
Types of Articles: Research.
Major Contents: Topics include: microwave techniques, components and instrumentation; microwave sources; integrated optics; arrays; aperture antennas; travelling-wave antennas; microwave network theory, optical-fibre communications.

218 IEE PROCEEDINGS PART I: SOLID STATE AND ELECTRON DEVICES

Institution of Electrical Engineers
P.O. Box 26
Hitchin, Herts. SG5 1SA
England

Editor: B. Dunkley
ISSN: 0143-7100 **Circulation:** 2,500 **Frequency:** BM
Style Sheet: Yes
Revised Theses or Diss.: Yes **Manuscript Copies:** 4
Review Period: 6 months (PR) **Lagtime:** 6 months
Types of Articles: Research, reviews.
Major Contents: Topics covered include: discrete and integrated device processes; VLSI call design and characterization; circuit/technology interaction, modelling and CAD; display devices, transducers and optical components; and novel electron devices.

219 IEEE ENGINEERING IN MEDICINE AND BIOLOGY MAGAZINE

Institute of Electrical and Electronics Engineers, Inc.
345 E. 47th Street
New York, NY 10017

Editor: Dr. Alvin S. Wald, Department of Anesthesiology, Columbia-Presbyterian Medical Center, 630 W. 168th Street, New York, NY 10032
Phone: 212-694-2575

ISSN: 0278-0054 **Circulation:** N/A **Frequency:** Q
Style Sheet: *See Miscellaneous
Revised Theses or Diss.: Yes **Manuscript Copies:** 4
Review Period: 3 months (PR) **Lagtime:** 9-12 months
Types of Articles: Research, communications, new product descriptions, and book reviews.
Major Contents: Instrumentation, research, and applications in biological and medical science.
Miscellaneous: *IEEE Power Engineering Society Publication Guide for Power Engineering Style Guide. It is IEEE's policy to own the copyright to technical papers it publishes. Authors must submit a completed IEEE Copyright Form with their manuscript.

220 IEEE ENGINEERING MANAGEMENT REVIEW
Institute of Electrical and Electronics Engineers, Inc.
345 E. 47th Street
New York, NY 10017

Editor: Irwin Gray, 175-14 73rd Avenue, Flusing, NY 11366
Phone: 212-969-0665
ISSN: 0360-8581 **Circulation:** N/A **Frequency:** Q
Style Sheet: *See Miscellaneous
Revised Theses or Diss.: Yes **Manuscript Copies:** 4
Review Period: 3 months (PR) **Lagtime:** 9-12 months
Types of Articles: Research and articles.
Major Contents: The theory or practice of engineering management. Papers are divided into three sections. "Beyond the Organization; the Nation and the World"; "Government and Business Function"; "Administrative Techniques."
Miscellaneous: *IEEE Power Engineering Society Publication Guide for Power Engineering Style Guide. It is IEEE's policy to own the copyright to technical papers it publishes. Authors must submit a completed IEEE Copyright Form with their manuscript prior to publication.

221 IEEE JOURNAL OF QUANTUM ELECTRONICS
Institute of Electrical and Electronics Engineers, Inc.
345 E. 47th Street
New York, NY 10017

Editor: Dr. William Streifer, Physics and Astronomy, Room 115, University of New Mexico, Albuquerque, NM 87131
Phone: 212-705-7900
ISSN: 0018-9197 **Circulation:** 3,000 **Frequency:** M
Style Sheet: *See Miscellaneous

Revised Theses or Diss.: Yes **Manuscript Copies:** 4
Review Period: 3 months (PR) **Lagtime:** 9-12 months
Types of Articles: Theory, design, research, and application papers, technical notes, and correspondence.
Major Contents: Quantum electronics and applications, including optics-electronic theory and techniques; lasers and fiber optics; design, development and manufacture of systems.
Miscellaneous: *IEEE Power Engineering Society Publication Guide for Power Engineering Style Guide. It is IEEE's policy to own the copyright to technical papers it publishes. Authors must submit a completed IEEE Copyright Form with their manuscript prior to publication.

222 **IEEE JOURNAL OF SOLID-STATE CIRCUITS**
Institute of Electrial and Electronics Engineers, Inc.
345 E. 47th Street
New York, NY 10017

Editor: W. David Pricer, NO1/965-2, General Technology Division, IBM Corporation, Essex Junction, VT 05452
Phone: 212-705-7900
ISSN: 0018-9200 **Circulation:** N/A **Frequency:** BM
Style Sheet: *See Miscellaneous
Revised Theses or Diss.: Yes **Manuscript Copies:** 4
Review Period: 3 months (PR) **Lagtime:** 9-12 months
Types of Articles: Research.
Major Contents: Device coverage in the areas of central importance to circuits, and system coverage in areas contiguous with circuit considerations.
Miscellaneous: *IEEE Power Engineering Society Publication Guide for Power Engineering Style Guide. It is IEEE's policy to own the copyright to technical papers it publishes. Authors must submit a completed IEEE Copyright Form with their manuscript prior to publication.

223 **IEEE SOFTWARE**
Institute of Electrical & Electronics Engineers Computer Society
345 E. 47th Street
New York NY 10017

Editor: Bruce D. Shriver, Editor-in-Chief, IBM, Thomas J. Watson Research Center, P.O. Box 218, Yorktown Heights, NY 10598
Phone: 714-821-8380
ISSN: 0704-7459 **Circulation:** 21,000 **Frequency:** Q
Style Sheet: *See Miscellaneous
Revised Theses or Diss.: Yes **Manuscript Copies:** 4

Review Period: 4-6 months (PR) **Lagtime:** 9-12 months
Types of Articles: Research, feature articles, book reviews, and new product reviews.
Major Contents: While not specifically related to energy applications, software problems, developments, testing and applications in any energy field would be appropriate.
Miscellaneous: *IEEE Power Engineering Society Publication Guide for Power Engineering Style Guide. Authors must submit a completed IEEE Copyright Form with their manuscript prior to publication.

224 **IEEE SPECTRUM**
Institute of Electrical and Electronics Engineers, Inc.
345 E. 47th Street
New York, NY 10017

Editor: Donald Christiansen
Phone: 212-644-7554
ISSN: 0018-9235 **Circulation:** 100,000 **Frequency:** M
Style Sheet: *See Miscellaneous
Revised Theses or Diss.: Yes **Manuscript Copies:** 3
Review Period: 2 months (PR) **Lagtime:** 3-4 months
Types of Articles: Research, projects, book reviews, and reviews of industry and professional news.
Major Contents: All areas of electronics and electrical engineering, including fiber optics, computers, satellites, power systems design and engineering, superconducting devices, electronic circuits and devices.
Miscellaneous: Formerly: ELECTRICAL ENGINEERING, ISSN: 0095-9197. *IEEE Power Engineering Society Publication Guide for Power Engineering.

225 **IEEE TECHNOLOGY AND SOCIETY MAGAZINE**
Institute of Electrical and Electronics Engineers, Inc.
345 E. 47th Street
New York, NY 10017

Editor: Dr. Normal Balabanian, Electrical Engineering Department, 111 Link Hall, Syracuse University, Syracuse, NY 13210
Phone: 212-705-7900
ISSN: 0278-0097 **Circulation:** N/A **Frequency:** Q
Style Sheet: *See Miscellaneous
Revised Theses or Diss.: Yes **Manuscript Copies:** 4
Review Period: 3 months (PR) **Lagtime:** 9-12 months
Types of Articles: Discursive, analytical, or tutorial articles, letters to the editor, and book reviews.

Major Contents: All aspects of social implications of technology.
Miscellaneous: *IEEE Power Engineering Society Publication Guide for Power Engineering Style Guide. It is IEEE's policy to own the copyright to technical papers it publishes. Authors must submit a completed IEEE Copyright Form with their manuscript prior to publication.

226 IEEE TRANSACTIONS ON ANTENNAS AND PROPAGATION

Institute of Electrical and Electronics Engineers, Inc.
345 E. 47th Street
New York, NY 10017

Editor: Ronald Fante, Avco Systems Division, 201 Lowell Street, Wilmington, MA 01887
Phone: 212-705-7900
ISSN: 0018-926X **Circulation:** 6,600 **Frequency:** M
Style Sheet: *See Miscellaneous
Revised Theses or Diss.: Yes **Manuscript Copies:** 4
Review Period: 3 months (PR) **Lagtime:** 9-12 months
Types of Articles: Theory, research, design and application papers, technical notes and correspondence.
Major Contents: Experimental and theoretical advances in electromagnetic theory and in the radiation propagation, scattering and diffraction of electromagnetic waves and devices.
Miscellaneous: *IEEE Power Engineering Society Publication Guide for Power Engineering Style Guide. It is IEEE's policy to own the copyright to technical papers it publishes. Authors must submit a completed IEEE Copyright Form with their manuscript prior to publication.

227 IEEE TRANSACTIONS ON AUTOMATIC CONTROL

Institute of Electrical and Electronics Engineers, Inc.
345 E. 47th Street
New York, NY 10017

Editor: Abraham Haddad, School of Electrical Engineering, Georgia Institute of Technology, Atlanta, GA 30332-0250
Phone: 404-894-3930
ISSN: 0018-9286 **Circulation:** 7,200 **Frequency:** M
Style Sheet: *See Miscellaneous
Revised Theses or Diss.: Yes **Manuscript Copies:** 7
Review Period: 3-4 months (PR) **Lagtime:** 9-12 months
Types of Articles: Theory, design, research and application papers, book reviews, technical notes, and correspondence.
Major Contents: Real-time control, optimal control, adaptive and stochastic control, estimation and identification, linear systems, system

modeling and applications of physical, economic, and social systems.

Miscellaneous: *IEEE Power Engineering Society Publication Guide for Power Engineering Style Guide. It is IEEE's policy to own the copyright to technical papers it publishes. Authors must submit a completed IEEE Copyright Form with their manuscript prior to publication.

228 IEEE TRANSACTIONS ON BIO-MEDICAL ENGINEERING

Institute of Electrical and Electronics Engineers, Inc.
345 E. 47th Street
New York, NY 10017

Editor: Dr. Donald G. Childers, Department of Electrical Engineering, University of Florida, Gainesville, FL 32611

Phone: 904-392-4914

ISSN: 0018-9294 **Circulation:** N/A **Frequency:** M

Style Sheet: *See Miscellaneous

Revised Theses or Diss.: Yes **Manuscript Copies:** 4

Review Period: 3 months (PR) **Lagtime:** 9-12 months

Types of Articles: Research papers, technical communications, and announcements.

Major Contents: All areas of biomedical engineering.

Miscellaneous: *IEEE Power Engineering Society Publication Guide for Power Engineering Style Guide. It is IEEE's policy to own the copyright to technical papers it publishes. Authors must submit a completed IEEE Copyright Form with their manuscript prior to publication.

229 IEEE TRANSACTIONS ON CIRCUITS AND SYSTEMS

Institute of Electrical and Electronics Engineers, Inc.
345 E. 47th Street
New York, NY 10017

Editor: Andreaus Antonious, Department of Electrical Engineering, University of Victoria, P.O. Box 1700, Victoria, B.C. V8W 2Y2, Canada

Phone: 212-705-7900

ISSN: 0098-4094 **Circulation:** 11,000 **Frequency:** M

Style Sheet: *See Miscellaneous

Revised Theses or Diss.: Yes **Manuscript Copies:** 4

Review Period: 3 months **Lagtime:** 9-12 months

Types of Articles: Research papers, short original technical results, brief letters, communicating, reporting new results and their significance.

Major Contents: Electronic circuits and systems theory.

Miscellaneous: *IEEE Power Engineering Society Publication Guide for Power Engineering Style Guide. Authors must submit a completed IEEE Copyright Form with their manuscript prior to publication.

230 IEEE TRANSACTIONS ON COMPONENTS, HYBRIDS, & MANUFACTURING TECHNOLOGY

Institute of Electrical and Electronics Engineers, Inc.
345 E. 47th Street
New York, NY 10017

Editor: Nicholas Panousis, IBM Corporation, 5600 Cattle Road, San Jose, CA 95193
Phone: 408-256-4208
ISSN: 0148-6411 **Circulation:** N/A **Frequency:** Q
Style Sheet: *See Miscellaneous
Revised Theses or Diss.: Yes **Manuscript Copies:** 4
Review Period: 3 months **Lagtime:** 9-12 months
Types of Articles: Theoretical, experimental and applied research, communications, and discussions on previously published papers.
Major Contents: Electronic components, circuits, transistors, capacitors, chips, etc.
Miscellaneous: *IEEE Power Engineering Society Publication Guide for Power Engineering Style Guide. It is IEEE's policy to own the copyright to technical papers it publishes. Authors must submit a completed IEEE Copyright Form with their manuscript prior to publication.

231 IEEE TRANSACTIONS ON ELECTRICAL INSULATIONS

Institute of Electrical and Electronics Engineers, Inc.
345 E. 47th Street
New York, NY 10017

Editor: A. van Roggen, R.D.2, Kenneth Square, PA 19348
Phone: 215-388-6909
ISSN: 0018-9367 **Circulation:** N/A **Frequency:** BM
Style Sheet: *See Miscellaneous
Revised Theses or Diss.: Yes **Manuscript Copies:** 4
Review Period: 3 months (PR) **Lagtime:** 9-12 months
Types of Articles: Theoretical and applied research, communications and discussions, and synoptic papers on a subject of general interest.
Major Contents: Dielectric phenomena and measurement, gaseous, liquid and solid electrical insulating materials, and uses of these dielectric materials in electric and electronic circuits and systems under all conditions of use.
Miscellaneous: *IEEE Power Engineering Society Publication Guide for Power Engineering Style Guide. It is IEEE's policy to own the copyright to technical papers it publishes. Authors must submit a completed IEEE Copyright Form with their manuscript prior to publication.

232 IEEE TRANSACTIONS ON ELECTROMAGNETIC COMPATIBILITY

Institute of Electrical and Electronics Engineers, Inc.
345 E. 47th Street
New York, NY 10017

Editor: R.B. Schulz, Xerox Corporation, IPD, 1301 Ridgeview Drive, M.S. 330, Lewisville, TX 75067
Phone: 214-420-7919
ISSN: 0018-9375 **Circulation:** N/A **Frequency:** Q
Style Sheet: *See Miscellaneous
Revised Theses or Diss.: Yes **Manuscript Copies:** 4
Review Period: 3 months (PR) **Lagtime:** 9-12 months
Types of Articles: Theoretical, experimental and state-of-the-art research, tutorial information, and comments on previously published papers.
Major Contents: All areas of electromagnetic compatibility including: instrumentation and methods of measurement; origins and characteristics of electromagnetic emissions; emissions and susceptibility of electronic systems, and subsystems; off-frequency/off-beam antennas; utilization of electromagnetic spectrum; consideration of nuclear pulses; radiation hazards; and walsh functions.
Miscellaneous: *IEEE Power Engineering Society Publication Guide for Power Engineering Style Guide. It is IEEE's policy to own the copyright to technical papers it publishes. Authors must submit a completed IEEE Copyright Form with their manuscript prior to publication.

233 IEEE TRANSACTIONS ON ELECTRON DEVICES

Institute of Electrical and Electronics Engineers, Inc.
345 E. 47th Street
New York, NY 10017

Editor: David L. Carter, Room 3B-319, Bell Laboratories, 600 Mountain Avenue, Murray Hill, NJ 07974
Phone: 212-705-7900
ISSN: 0018-9383 **Circulation:** N/A **Frequency:** M
Style Sheet: *See Miscellaneous
Revised Theses or Diss.: Yes **Manuscript Copies:** 4
Review Period: 3 months (PR) **Lagtime:** 9-12 months
Types of Articles: Research, tutorial, reviews, and comments on previously published papers.
Major Contents: Theory, design, and performance of active electron and ion devices, solid-state devices, integrated electronic devices, and en-

ergy sources.

Miscellaneous: *IEEE Power Engineering Society Publication Guide for Power Engineering Style Guide. It is IEEE's policy to own the copyright to technical papers it publishes. Authors must submit a completed IEEE Copyright Form with their manuscript prior to publication.

234 IEEE TRANSACTIONS ON GEOSCIENCE AND REMOTE SENSING

Institute of Electrical and Electronics Engineers, Inc.
345 E. 47th Street
New York, NY 10017

Editor: Fawwaz Ulaby, University of Michigan Radiation Laboratory, Department of Electrical & Computer Engineer., 4072 E. Engineering Building, Ann Arbor, MI 47109
Phone: 212-705-7900
ISSN: 0196-2892 **Circulation:** N/A **Frequency:** BM
Style Sheet: *See Miscellaneous
Revised Theses or Diss.: Yes **Manuscript Copies:** 5
Review Period: 3 months (PR) **Lagtime:** 9-12 months
Types of Articles: Research, reviews, tutorials, brief papers, and letters ot the editor.
Major Contents: The theory, concepts, and techniques of science and engineering as applied to sensing the earth, oceans, atmosphere, and space, and the processing, interpretation, and dissemination of this information.
Miscellaneous: *IEEE Power Engineering Society Publication Guide for Power Engineering Style Guide. It is IEEE's policy to own the copyright to technical papers it publishes. Authors must submit a completed IEEE Copyright Form with their manuscript prior to publication.

235 IEEE TRANSACTIONS ON INDUSTRIAL ELECTRONICS AND CONTROL INSTRUMENTATION

Institute of Electrical & Electronics Engineers
345 E. 47th Street
New York, NY 10017

Editor: Gerald Cook, Department of Electrical & Computer Engineering, George Mason University, Fairfax, VA 22030
Phone: 703-323-2302
ISSN: 0278-0046 **Circulation:** N/A **Frequency:** Q
Style Sheet: See Miscellaneous
Revised Theses or Diss.: N/A **Manuscript Copies:** 4

Review Period: 3 months **Lagtime:** 9-12 months
Types of Articles: Research, surveys, and comments on previously published papers.
Major Contents: All aspects of the theory and application of industrial electronics and control instrumentation science and engineering, including microprocessor control systems, high power controls, process control, programmable controllers, numerical and program control systems, flow meters, and identification systems.
Miscellaneous: IEEE Power Engineering Society Publication Guide for Power Engineering Style Guide. It is IEEE's policy to own the copyright to technical papers it publishes. Authors must submit a completed IEEE Copyright Form with their manuscript prior to publication.

236 IEEE TRANSACTIONS ON INDUSTRY APPLICATIONS

Institute of Electrical and Electronic Engineers, Inc.
345 E. 47th Street
New York, NY 10017

Editor: A. Edward Rich, IEEE Transactions on Industry Applications, 243 Juniper Drive, Schenectady, NY 12306
Phone: 212-705-7900
ISSN: 0093-9994 **Circulation:** N/A **Frequency:** BM
Style Sheet: *See Miscellaneous
Revised Theses or Diss.: Yes **Manuscript Copies:** 4
Review Period: 3 months (PR) **Lagtime:** 9-12 months
Types of Articles: Research, novel applications, state-of-the-art tutorials, and comments on previously published articles.
Major Contents: Research and application studies relating to the design, installation, operation and maintenance of electrical systems, equipment, and devices for safe, reliable, and economical performance.
Miscellaneous: *IEEE Power Engineering Society Publication Guide for Power Engineering Style Guide. Authors must submit a completed IEEE Copyright Form with their manuscript prior to publication.

237 IEEE TRANSACTIONS ON MAGNETICS

Institute of Electrical and Electronics Engineers, Inc.
345 E. 47th Street
New York, NY 10017

Editor: Stanley H. Charap, Department Electrical and Computer Engineering, Carnegie-Mellon University, Schenley Park, Pittsburgh, PA 15213
Phone: 212-705-7900

ISSN: 0018-9464 **Circulation:** N/A **Frequency:** BM
Style Sheet: *See Miscellaneous
Revised Theses or Diss.: Yes **Manuscript Copies:** 4
Review Period: 3 months **Lagtime:** 9-12 months
Types of Articles: Technical papers, literature surveys and critical reviews, computer program descriptions; conference papers, and short technical communications.
Major Contents: Treats the fundamental development, design, and application of magnetic devices as well as magnetic materials and phenomena as relevant to engineering purpose. Superconductivity papers on high magnetic field generation and switching are also solicited.
Miscellaneous: *IEEE Power Engineering Society Publication Guide for Power Engineering Style Guide. It is IEEE's policy to won the copyright to technical papers it publishes. Authors must submit a completed IEEE Copyright Form with their manuscript prior to publication.

238 IEEE TRANSACTIONS ON MICROWAVE THEORY AND TECHNIQUES

Institute of Electrical and Electronics Engineers, Inc.
345 E. 47th Street
New York, NY 10017

Editor: Dr. Tatsuo Itoh, Department of Electrical Engineering, University of Texas at Austin, Austin, TX 78712
Phone: 212-705-7900
ISSN: 0018-9480 **Circulation:** N/A **Frequency:** M
Style Sheet: *See Miscellaneous
Revised Theses or Diss.: Yes **Manuscript Copies:** 4
Review Period: 3 months (PR) **Lagtime:** 9-12 months
Types of Articles: Research, short papers, and letters.
Major Contents: Microwave theory, techniques, and applications.
Miscellaneous: *IEEE Power Engineering Society Publication Guide for Power Engineering Style Guide. It is IEEE's policy to own the copyright to technical papers it publishes. Authors must submit a completed IEEE Copyright Form with their manuscript prior to publication.

239 IEEE TRANSACTIONS ON NUCLEAR SCIENCE

Institute of Electrical and Electronic Engineers
345 E. 47th Street
New York, NY 10017

Editor: R.F. Shea
Phone: 212-705-7555

ISSN: 0018-9499 **Circulation:** N/A **Frequency:** BM
Style Sheet: *See Miscellaneous
Revised Theses or Diss.: N/A **Manuscript Copies:** 3
Review Period: 2-3 months (PR) **Lagtime:** N/A
Types of Articles: Research, tutorials, and correspondence.
Major Contents: All aspects of theory and application of nuclear science, including: instrumentation, engineering, high-energy physics, reactor controls, and radiation effects.
Miscellaneous: Submit original and final manuscript to R.F. Shea, 709 Cypress Place, Sun City Center, FL 33570, (May 1-Oct. 31), and Nov. 1-April 30) R.F. Shea, 7 Barkley Street, South Yarmouth, MA 02664. *IEEE Power Engineering Society Publication Guide for Power Engineering Style Guide.

240 IEEE TRANSACTIONS ON POWER APPARATUS AND SYSTEMS
Institute of Electrical and Electronic Engineers, Inc.
345 E. 47th Street
New York, NY 10017

Editor: C.J. Essel, 5969 W. 76th Street, Los Angeles, CA 90045
Phone: 213-645-3380
ISSN: 0018-9510 **Circulation:** N/A **Frequency:** M
Style Sheet: *See Miscellaneous
Revised Theses or Diss.: N/A **Manuscript Copies:** N/A
Review Period: 2 months (PR) **Lagtime:** 3-4 months
Types of Articles: Research.
Major Contents: All aspects of the science and practice of electric power generation, transmission, distribution, and utilization.
Miscellaneous: One page summaries of papers published in full are printed in IEEE POWER ENGINEERING REVIEW. *IEEE Power Engineering Society Publication Guide for Power Engineering.

241 IEEE TRANSACTIONS ON SOFTWARE ENGINEERING
Institute of Electrical and Electronics Engineers, Inc.
345 E. 47th Street
New York, NY 10017

Editor: Raymond T. Yeh, Editor-in-Chief, Department of Computer Science, University of Maryland, College Park, MD 20742
Phone: 212-705-7900
ISSN: 0098-5584 **Circulation:** N/A **Frequency:** BM
Style Sheet: *See Miscellaneous
Revised Theses or Diss.: Yes **Manuscript Copies:** 4

Review Period: 3 months (PR) **Lagtime:** 9-12 months
Types of Articles: Research and correspondence.
Major Contents: The scope of this journal will cover all areas which form the middle ground that lies between the initial step of basic research and the end use of computer software. More specifically, it includes the following areas: requirement analysis and specification, programming methodology, software testing and validation, performance and design evaluations, software project management, and programming tools and standards.
Miscellaneous: *IEEE Power Engineering Society Publication Guide for Power Engineering Style Guide. It is IEEE's policy to own the copyright to technical papers it publishes. Authors must submit a completed IEEE Copyright Form with their manuscript prior to publication.

242 IEEE TRANSACTIONS ON SONICS AND ULTRASONICS

Institute of Electrical and Electronics Engineers, Inc.
345 E. 47th Street
New York, NY 10017

Editor: S. Wanuga, General Electric Company, Electronics Lab 244, Electronics Park, Syracuse, NY 13221
Phone: 212-705-7900
ISSN: 0018-9537 **Circulation:** N/A **Frequency:** BM
Style Sheet: *See Miscellaneous
Revised Theses or Diss.: Yes **Manuscript Copies:** 4
Review Period: 3 months (PR) **Lagtime:** 9-12 months
Types of Articles: Research, device applications, and book reviews.
Major Contents: New research and application results in the field of sonics and ultrasonics.
Miscellaneous: *IEEE Power Engineering Society Publication Guide for Power Engineering Style Guide. It is IEEE's policy to own the copyright to technical papers it publishes. Authors must submit a completed IEEE Copyright Form with their manuscript prior to publication.

243 IEEE TRANSACTIONS ON SYSTEMS, MAN, AND CYBERNETICS

Institute of Electrical and Electronics Engineers, Inc.
345 E. 47th Street
New York, NY 10017

Editor: Andrew P. Sage, George Mason University, 4400 University Drive, Fairfax, VA 22030
Phone: 703-323-2939

ISSN: 0018-9472 **Circulation:** N/A **Frequency:** BM
Style Sheet: *See Miscellaneous
Revised Theses or Diss.: Yes **Manuscript Copies:** 4
Review Period: 3 months (PR) **Lagtime:** 9-12 months
Types of Articles: Research, technical papers, short technical, and communications.
Major Contents: Large-scale systems, theory and applications; optimization, decision analysis, problem definition, modeling, simulation, test, and evaluation. Foundations of cybernetics, pattern recognition, adaptive and learning systems, biocybernetics; man-machine systems. Representative applications include complex hardware, behavioral, biological, ecological, educational, environmental health care, management, socioeconomic, transportation, and urban systems; national priorities.
Miscellaneous: *IEEE Power Engineering Society Publication Guide for Power Engineering Style Guide. It is IEEE's policy to own the copyright to technical papers it publishes. Authors must submit a completed IEEE Copyright Form with their manuscript prior to publication.

244 IEEE TRANSACTIONS ON VEHICULAR TECHNOLOGY
Institute of Electrical and Electronics Engineers, Inc.
345 E. 47th Street
New York, NY 10017

Editor: George F. McClure, Martin Marietta Aerospace, Sand Lake Road, Box 5837, Mail Point 362, Orlando, FL 32855
Phone: 305-352-3782
ISSN: 0018-9545 **Circulation:** N/A **Frequency:** Q
Style Sheet: *See Miscellaneous
Revised Theses or Diss.: Yes **Manuscript Copies:** 4
Review Period: 3 months (PR) **Lagtime:** 9-12 months
Types of Articles: Theory, research, design, and application papers, technical notes, and correspondence.
Major Contents: Electrical and electronics technology in vehicles and vehicular systems, including: traffic control systems, vehicle communications systems, and automotive technology.
Miscellaneous: *IEEE Power Engineering Society Publication Guide for Power Engineering Style Guide. It is IEEE's policy to own the copyright to technical papers it publishes. Authors must submit a completed IEEE Copyright Form with their manuscript prior to publication.

245 ITE JOURNAL
Institute of Transportation Engineers
525 School Street, N.W.
Washington, DC 20004

Editor: Nancy D. Angarola
Phone: 202-554-8050
ISSN: 0148-0170 **Circulation:** M **Frequency:** 8,100
Style Sheet: Yes
Revised Theses or Diss.: No **Manuscript Copies:** 4
Review Period: 1-2 months **Lagtime:** 3-6 months
Types of Articles: Research, news items, letters to the editor, book reviews, and informative articles.
Major Contents: Covers interesting and new developments in transportation and traffic management, especially ground transportation. Topic could range from the transportation of people to the safe transport of hazardous material and from lane markers to computerized traffic systems.
Miscellaneous: For engineers responsible for planning, design and operation of street and highway systems. An official Journal of the Institute of Transportation Engineers.

246 **I.U.G.G. CHRONICLE**

International Union of Geodesy and Geophysics
39 Ter Rue Gay-Lussac
75005 Paris, France

Editor: Paul Melchior and Georges Laclavere
Phone: 33-1-354 Ext. 21
ISSN: 0047-1259 **Circulation:** 1,400 **Frequency:** BM
Style Sheet: No
Revised Theses or Diss.: No **Manuscript Copies:** 2
Review Period: 1-2 months (PR) **Lagtime:** 2-3 months
Types of Articles: Announcements and reviews of symposia, seminars, and conferences. Description and reports on large interdisciplinary and international programs; occasionally scientific papers.
Major Contents: Geophysics and tectonophysics; planetology announcement and reports on scientific meetings of geophysical interest.

247 **IN SITU: OIL. COAL. SHALE. MINERALS**

Marcel Dekker, Inc.
270 Madison Avenue
New York, NY 10016

Editor: T.F. Edgar, Department of Chemical Engineering, University of Texas, Austin, TX 78712
Phone: 512-471-3080

ISSN: 0146-2520 **Circulation:** N/A **Frequency:** Q
Style Sheet: Yes
Revised Theses or Diss.: Yes **Manuscript Copies:** 5
Review Period: 3 months (PR) **Lagtime:** 6 months
Types of Articles: Research.
Major Contents: In Situ processing of minerals and fossil fuels, especially in the emerging technologies relating to oil, coal, shale, tar sands, uranium, salt, copper, and other minerals. Underground storage of these materials and wastes are also considered.

248 **IN TECH**
Instrument Society of America
ICON, 211 S. 45th Street
Philadelphia, PA 19104

Editor: Alan Krigman, Editorial Director
Phone: 919-549-8411
ISSN: 0192-303X **Circulation:** 40,000 **Frequency:** M
Style Sheet: Yes
Revised Theses or Diss.: Yes **Manuscript Copies:** 1
Review Period: 1-2 months **Lagtime:** 3 months
Types of Articles: Research, projects, and reviews.
Major Contents: Focuses on instrumentation technology. Application-oriented articles relating to monitoring and control of industrial processes. Typical articles are of interest to production people in chemical, petrochemical, petroleum, pulp and paper, textiles, utility, food and beverage, pharmaceuticals, and rubber and plastics industries.
Miscellaneous: Formerly: ISA JOURNAL (1954-1966), INSTRUMENTATION TECHNOLOGY.

249 **INDUSTRIAL AND ENGINEERING CHEMISTRY FUNDAMENTALS**
American Chemical Society
1155 16th Street, N.W.
Washington, DC 20036

Editor: Dr. Robert L. Pigford, Department of Chemical Engineering, University of Delaware, Newark, DE 19711
Phone: 202-452-8723
ISSN: 0196-4313 **Circulation:** 4,600 **Frequency:** Q
Style Sheet: *See Miscellaneous
Revised Theses or Diss.: Yes **Manuscript Copies:** 4
Review Period: 8-10 months (PR) **Lagtime:** 3-4 months
Types of Articles: Research, project reports, short communications, and

reviews.
Major Contents: All aspects of chemical engineering research.
Miscellaneous: Author's guide is published in the first issue of the year. *American Chemical Society Handbook for Authors.

250 **INGENIERIA QUIMICA**
Triana, 51
28016 Madrid, Spain

Editor: Pedro de la Pezuela
Phone: 341-457-6400
ISSN: 0210-2064 **Circulation:** 7,000 **Frequency:** M
Style Sheet: No
Revised Theses or Diss.: Yes **Manuscript Copies:** 1
Review Period: 4 months **Lagtime:** 3-6 months
Types of Articles: Research, projects, reviews.
Major Contents: General and specific articles on chemical plants and chemical engineering especially design, engineering, construction, operation and maintenance of chemical plants.

251 **INSTITUTE OF INDUSTRIAL ENGINEERS TRANSACTIONS (IIE)**
Institute of Industrial Engineers
25 Technology Park/Atlanta.
Norcross, GA 30092

Editor: Thom J. Hodgson, Industrial Engineering Department, North Carolina State University, Raleigh, NC 27695-7906
Phone: 919-737-2362
ISSN: 0740-817X **Circulation:** N/A **Frequency:** Q
Style Sheet: Yes
Revised Theses or Diss.: N/A **Manuscript Copies:** 4
Review Period: Varies (PR) **Lagtime:** 6-24 months
Types of Articles: Research, surveys, book reviews, and technical notes.
Major Contents: New developments and research in industrial engineering such as applied optimization; applied probability and statistics; development and implementation of manufacturing systems; production and inventory systems; facilities design; scheduling; planning; simulation; computer and information sciences.

252 **INTERNATIONAL BIODETERIORATION**
Commonwealth Mycological Institute
Senior Consultant Biodeterioration Center
Ferry Lane, Kew
Surrey TW9 3AF, England

Editor: H.O.W. Eggins, Editor-in-Chief
Phone: 01-940-4086
ISSN: 0265-3036 **Circulation:** N/A **Frequency:** Q
Style Sheet: Yes
Revised Theses or Diss.: Yes **Manuscript Copies:** 2
Review Period: N/A (PR) **Lagtime:** N/A
Types of Articles: Research, projects, reviews, news and abstracts.
Major Contents: All aspects of biodeterioration and biodegradation.
Miscellaneous: Incorporating INTERNATIONAL BIODETERIORATION BULLETIN; BIODETERIORATION RESEARCH TITLES; and WASTE MATERIALS BIODEGRADATION TITLES.

253 INTERNATIONAL COMMUNICATIONS IN HEAT AND MASS TRANSFER

Pergamon Press
Maxwell House, Fairview Park
Elmsford, NY 10523

Editor: J.P. Hartnett and W.J. Minkwycz
Phone: 914-592-7000
ISSN: 0735-1933 **Circulation:** 1,050 **Frequency:** BM
Style Sheet: Yes
Revised Theses or Diss.: Yes **Manuscript Copies:** 3
Review Period: 1 month **Lagtime:** 2 months
Types of Articles: Research and letters.
Major Contents: New ideas in the field of heat and mass transfer.
Miscellaneous: Serves as a springboard for communication and discussion of new ideas in the field. Formerly: LETTERS IN HEAT AND MASS TRANSFER, ISSN: 0094-4548.

254 INTERNATIONAL JOURNAL FOR HOUSING SCIENCE AND ITS APPLICATIONS

Pergamon Press
Maxwell House, Fairview Park
Elmsford, NY 10523

Editor: Professor Oktay Ural, International Association for Housing Science, P.O. Box 340254, Coral Gables, FL 33134
Phone: 914-592-7700
ISSN: 0146-6518 **Circulation:** 1,000 **Frequency:** BM
Style Sheet: Yes
Revised Theses or Diss.: N/A **Manuscript Copies:** N/A

Review Period: 1 month (PR) **Lagtime:** 12 months
Types of Articles: Research and correspondence.
Major Contents: Planning, design, construction, financing, marketing, and management of housing.

255 INTERNATIONAL JOURNAL FOR NUMERICAL AND ANALYTICAL METHODS IN GEOMECHANICS

John Wiley and Sons, Ltd.
Baffins Lane
Chichester, Sussex, England

Editor: Professor C.S. Desai, Department of Civil Engineering and, Engineering Mechanics, University of Arizona, Tucson, AZ 85721
Phone: 602-621-6569
ISSN: 0363-9061 **Circulation:** 700 **Frequency:** BM
Style Sheet: Yes
Revised Theses or Diss.: No **Manuscript Copies:** 4
Review Period: 1-3 months (PR) **Lagtime:** 8-10 months
Types of Articles: Research, projects, and reviews of interdisciplinary subjects covering theory, formulations and applications of numerical methods, and constitutive laws useful to the researcher and practitioner.
Major Contents: Soil and rock mechanics, mechanics of structures and foundations, earth structure interaction, offshore and marine technology, oil and fluid flow and diffusion, geological engineering, geothermal energy, and ice mechanics.

256 INTERNATIONAL JOURNAL FOR NUMERICAL METHODS IN ENGINEERING

John Wiley and Sons, Inc.
605 3rd Avenue
New York, NY 10158

Editor: Olgierd C. Zienkiewicz
Phone: 212-850-6416
ISSN: 0029-5981 **Circulation:** N/A **Frequency:** M
Style Sheet: Yes
Revised Theses or Diss.: N/A **Manuscript Copies:** 3
Review Period: 3-4 months (PR) **Lagtime:** 12 months
Types of Articles: Research, computer papers, and letters to editors.
Major Contents: Structural analysis, heat transfer, fluid mechanics and optimal system design, also computer-aided design activity.

257 **INTERNATIONAL JOURNAL OF APPLIED RADIATION AND ISOTOPES**

Pergamon Press
Maxwell House, Fairview Park
Elmsford, NY 10523

Editor: Wilfred B. Mann
Phone: 914-592-7700
ISSN: 0020-708X **Circulation:** 2,000 **Frequency:** M
Style Sheet: Yes
Revised Theses or Diss.: N/A **Manuscript Copies:** 2
Review Period: 3 weeks (PR) **Lagtime:** 7 months
Types of Articles: Research, technical notes, reviews, and letters to the editor.
Major Contents: Research or technological results of using radiation and radioactivity.

258 **INTERNATIONAL JOURNAL OF BIOMETEOROLOGY**

Swets Publishing Service
Heereweg 347B
2161 CA Lisse
The Netherlands

Editor: R.W. Gloyne, Editor-in-Chief, International Journal of Biometeorology, 2, Canowie Road, Redland, Bristol BS6 7HS, Great Britain
Phone: 44-272-730-996
ISSN: 0200-7128 **Circulation:** 1,300 **Frequency:** Q
Style Sheet: Yes
Revised Theses or Diss.: Yes **Manuscript Copies:** 2
Review Period: 3-9 months (PR) **Lagtime:** 6 months
Types of Articles: Research papers, provided "research" is liberally interpreted; occasional reviews; summaries of previously published work; short reports on research in progress; and book reviews.
Major Contents: Effects of meteorological elements on man, animals and plants; aerobiology; transport of spores; pollutants...by wind energy from biomass; and solar and wind energy.

259 **INTERNATIONAL JOURNAL OF COAL GEOLOGY**

Elsevier Science Publishers B.V.
P.O. Box 211
1000 AE Amsterdam
The Netherlands

Editor: W. Spackman, Editor-in-Chief, Coal Research Section, Pennsylvania State University, University Park, PA 16802
Phone: 020-5803-911
ISSN: 0166-5162 **Circulation:** N/A **Frequency:** Q
Style Sheet: Yes
Revised Theses or Diss.: N/A **Manuscript Copies:** 3
Review Period: Varies **Lagtime:** Varies
Types of Articles: Symposia, proceedings, surveys, book reviews, overviews of recent literature, and letters to the editor.
Major Contents: This journal includes the genesis of coal and coal seams, including studies of modern coal-forming processes and environments; the metamorphosis of coal and coal seams and coal dispersed in other rock types; the geology of coal measures including stratigraphic, structural, geomorphic, paleogeographic, hydrogeologic, palynologic, paleozoologic and paleobotanic facets; the petrology and petrography of coal and coal seams including coal mineralogy, geochemistry and the chemical and physical constitution of coal.
Miscellaneous: A detailed "Guide to Authors" is available on request.

260 INTERNATIONAL JOURNAL OF ENERGY RESOURCES

John Wiley and Sons, Ltd.
Baffins Lane, Chichester
Sussex PO1 IUD
England

Editor: J.T. McMullen, The New University of Ulster, Coleraine BT52 1SA, Northern Ireland
Phone: 212-850-6418
ISSN: N/A **Circulation:** N/A **Frequency:** Q
Style Sheet: Yes
Revised Theses or Diss.: N/A **Manuscript Copies:** N/A
Review Period: N/A **Lagtime:** N/A
Types of Articles: Research, reviews, short communications, letters, and book reviews.
Major Contents: Development and exploitation of both traditional and new fuels and other sources of energy.

261 INTERNATIONAL JOURNAL OF ENERGY SYSTEMS

Acta Press
P.O. Box 25, Station G
Calgary, Alberta
T3A 2G1, Canada

Editor: B.L. Capehart, P.O. Box 3243, Station B, Calgary, Alberta, T2M 4L8, Canada
Phone: 403-270-3616
ISSN: 0226-1472 **Circulation:** 1,000 **Frequency:** 5/year
Style Sheet: Yes
Revised Theses or Diss.: Yes **Manuscript Copies:** 3
Review Period: 1 month (PR) **Lagtime:** 6-9 months
Types of Articles: Surveys, tutorials, short communications, and research.
Major Contents: Policy; planning; economics; simulation; modelling; decision; technology; alternate energy sources; energy and power systems.
Miscellaneous: An official publication of the International Association of Science and Technology for Development - IASTED.

262 INTERNATIONAL JOURNAL OF ENVIRONMENTAL STUDIES, THE

Gordon and Breach Science Publishers, Ltd.
1 Bedford Street
London WC2E 9TP
England

Editor: Professor J. Rose, 95, St. Andrews Road, N., St. Annes-On-Sea, FY8 2IG Lancaster, England
Phone: 0253-712099
ISSN: 0020-7233 **Circulation:** N/A **Frequency:** M
Style Sheet: Yes
Revised Theses or Diss.: Yes **Manuscript Copies:** N/A
Review Period: N/A **Lagtime:** N/A
Types of Articles: Scientific papers, book reviews, news of conferences, and research.
Major Contents: All aspects of environmental studies, and appeals to social, natural and applied scientists, as well as architects, town planners and other professional people.
Miscellaneous: Gordon and Breach Publishers in New York, Phone: 212-206-8900.

263 INTERNATIONAL JOURNAL OF FRACTURE

Sijthoff and Noordhoff International Publishers
P.O. Box 4
Alphen aan den Rijn
The Netherlands

Editor: Professor M.L. Williams

ISSN: 0376-9429 **Circulation:** 1,350 **Frequency:** M
Style Sheet: Yes
Revised Theses or Diss.: N/A **Manuscript Copies:** 3
Review Period: 2 months (PR) **Lagtime:** 12 months
Types of Articles: Research, reviews, technical notes, and reports.
Major Contents: Mechanisms that cause micro and macro fracture in all materials and the effect on engineering.

264 INTERNATIONAL JOURNAL OF FUSION ENERGY

Fusion Energy Foundation
Box 1438, Radio City Street
New York, NY 10101

Editor: Majorie Hecht, Managing Editor, Dr. Robert J. Moon, Editor-in-Chief, University of Chicago, 5801 S. Ellis Avenue, Chicago, IL 60637
Phone: 212-247-8439
ISSN: 0146-4981 **Circulation:** 5,000 **Frequency:** Q
Style Sheet: Yes
Revised Theses or Diss.: Yes **Manuscript Copies:** 3
Review Period: 2-3 months **Lagtime:** Varies
Types of Articles: Research, projects, and reviews.
Major Contents: Nuclear fusion and fission; directed energy processes; and biology, especially areas whose physics are of the same character as fusion and directed energy processes. Highly organized plasmas; directed energy beams.
Miscellaneous: This journal's policy is decidedly anti-peer review because "It is essential that articles in the journal are of such a nature that they bring about the birth of new concepts. That would be very unlikely to occur with the referee system of judging papers for acceptance that is now in use in most scientific journals." Dr. Robert J. Moon, Editor-in-Chief.

265 INTERNATIONAL JOURNAL OF HEAT AND MASS TRANSFER

Pergamon Press
Maxwell House, Fairview Park
Elmsford, NY 10523

Editor: Profs. J.P. Hartnett and W.J. Minkowycz, University of Illinois at Chicago, Department of Mechanical Engineering, P.O. Box 4348, Chicago, IL 60680
Phone: 312-996-4490/3467
ISSN: 0017-9310 **Circulation:** 2,500 **Frequency:** M
Style Sheet: Yes

Revised Theses or Diss.: No **Manuscript Copies:** 3
Review Period: 2-3 months (PR) **Lagtime:** 6 months
Types of Articles: Research.
Major Contents: Heat transfer, mass transfer, fluid mechanics, and related fields.

266 INTERNATIONAL JOURNAL OF HYDROGEN ENERGY

Pergamon Press
Maxwell House, Fairview Park
Elmsford, NY 10523

Editor: T. NeJat Veziroglu, International Association for Hydrogen Energy, P.O. Box 248266, Coral Gables, FL 33124
Phone: 305-284-4666
ISSN: 0360-3199 **Circulation:** 2,250 **Frequency:** M
Style Sheet: Yes
Revised Theses or Diss.: No **Manuscript Copies:** 3
Review Period: 1-2 months (PR) **Lagtime:** 6 months
Types of Articles: Research, reviews, and projects.
Major Contents: Hydrogen introduction; hydrogen transmission; hydrogen storage; hydrides and their applications; hydrogen distribution; hydrogen utilization; economical and environmental considerations; and safety.
Miscellaneous: An international bibliography of recent publications in hydrogen energy listed by title, author, and source appears regularly in the Journal. Serves as an official journal of the International Association for Hydrogen Energy.

267 INTERNATIONAL JOURNAL OF INFRARED AND MILLIMETER WAVES

Plenum Publishing Corporation
233 Spring Street
New York, NY 10013

Editor: Kenneth J. Button
Phone: 212-620-8000
ISSN: 0195-9271 **Circulation:** N/A **Frequency:** BM
Style Sheet: AIP Style Manual.
Revised Theses or Diss.: N/A **Manuscript Copies:** 2
Review Period: 2 weeks (PR) **Lagtime:** 2 months
Types of Articles: Original research and proceedings of the annual conference on infrared and millimeter waves.
Major Contents: Millimeter and far infrared theory, techniques, devices, systems, spectroscopy, and applications.

268 INTERNATIONAL JOURNAL OF MECHANICAL SCIENCES

Pergamon Press
Maxwell House, Fairview Park
Elmsford, NY 10523

Editor: W. Johnson, Editor-in-Chief, Ridge Hall, Chapel-en-le-Frith, Via Stockport, Cheshire SK12 6UD, England
Phone: 914-592-7700
ISSN: 0020-7403 **Circulation:** 1,400 **Frequency:** M
Style Sheet: Yes
Revised Theses or Diss.: N/A **Manuscript Copies:** 2
Review Period: 3 months (PR) **Lagtime:** 9-12 months
Types of Articles: Original research papers and book reviews.
Major Contents: Papers should concern the mechanical and civil engineering sciences or engineering practices.

269 INTERNATIONAL JOURNAL OF MINERAL PROCESSING

Elsevier Science Publishers B.V.
P.O. Box 330
1000 AH Amsterdam
The Netherlands

Editor: P. Gy and D. Fuerstenau
ISSN: 0301-7516 **Circulation:** N/A **Frequency:** Q
Style Sheet: Yes
Revised Theses or Diss.: N/A **Manuscript Copies:** 3
Review Period: 1 month (PR) **Lagtime:** 10 months
Types of Articles: Research, reviews, case histories, and book reviews.
Major Contents: Topics on the processing of solid mineral materials, such as: metallic and non-metallic ores, coal, and other solid sources of secondary materials. Preferred topics include: comminution; sizing; classification; gravity concentration; flotation; electric and magnetic separation; miscellaneous physical separation techniques; pyrometallurgy and hydrometallurgy; biometallurgy; biohydrometallurgy; waste treatment and disposal; and mineralogical studies.
Miscellaneous: "Guide for Authors" is available on request, and also appears in the first issue to appear each year.

270 INTERNATIONAL JOURNAL OF MINI AND MICROCOMPUTERS

Acta Press
P.O. Box 25, Station G
Calgary, Alberta
T3A 2G1, Canada

Editor: B. Furht
Phone: 403-286-1589
ISSN: 0702-0481 **Circulation:** 2,000 **Frequency:** 3/year
Style Sheet: No
Revised Theses or Diss.: Yes **Manuscript Copies:** 3
Review Period: 2 months (PR) **Lagtime:** N/A
Types of Articles: Research, book reviews, conferences, call for papers, new publications, and advertisements.
Major Contents: All aspects of mini and microcomputers including technology, hardware, software, systems, education, networks, distributed processing, and applications.
Miscellaneous: An official publication of The International Society for Mini and Microcomputers - ISMM.

271 INTERNATIONAL JOURNAL OF MODELLING AND SIMULATION

Acta Press
P.O. Box 25, Station G
Calgary, Alberta
T3A 2G1, Canada

Editor: S.G. Tzafestas
Phone: 403-270-3616
ISSN: 0228-6203 **Circulation:** 1,000 **Frequency:** Q
Style Sheet: No
Revised Theses or Diss.: Yes **Manuscript Copies:** 3
Review Period: 2 months (PR) **Lagtime:** N/A
Types of Articles: Research, tutorials, surveys and applications, book reviews, conferences, call for papers, new publications, and advertisements.
Major Contents: Modelling, simulation, languages, software, hardware, methodology, statistical techniques, numerical and graphical methods.

272 INTERNATIONAL JOURNAL OF MULTIPHASE FLOW

Pergamon Press
Maxwell House, Fairview Park
Elmsford, NY 10523

Editor: G. Hetsroni, Department of Mechanical Engineering, Technion, Haifa 32000, Israel
Phone: 914-592-7700
ISSN: 0301-9322 **Circulation:** 1,000 **Frequency:** BM
Style Sheet: Yes

Revised Theses or Diss.: Yes **Manuscript Copies:** 3
Review Period: 1 month (PR) **Lagtime:** 6 months
Types of Articles: Research, brief communications, and reviews.
Major Contents: Theoretical and experimental papers on multiphase flow (solid-fluid, fluid-fluid, etc.)
Miscellaneous: Gives new patents and patent application from U.S. and foreign countries.

273 INTERNATIONAL JOURNAL OF NON-LINEAR MECHANICS

Pergamon Press, Ltd.
Headington Hill Hall
Oxford, OX3 OBW
England

Editor: William A. Nash
Phone: 914-592-7700
ISSN: 0020-7462 **Circulation:** 1,200 **Frequency:** BM
Style Sheet: Yes
Revised Theses or Diss.: N/A **Manuscript Copies:** 2
Review Period: 4-6 months (PR) **Lagtime:** 12 months
Types of Articles: Theoretical and experimental papers.
Major Contents: Mechanics of solids and fluids, and control theory on non-linear phenomena.

274 INTERNATIONAL JOURNAL OF POLYMERIC MATERIALS

Gordon and Breach Science Publishers, Inc.
1 Park Avenue
New York, NY 10016

Editor: Herman S. Kaufman
Phone: 212-692-6000
ISSN: 0091-4037 **Circulation:** 2,065 **Frequency:** Q
Style Sheet: Yes
Revised Theses or Diss.: N/A **Manuscript Copies:** 3
Review Period: 1-2 months (PR) **Lagtime:** 6-9 months
Types of Articles: Research.
Major Contents: Designed to help the designer, fabricator, and consumer to make the best choice and use of polymers, especially as engineering materials.

275 INTERNATIONAL JOURNAL OF RADIATION, ONCOLOGY, BIOLOGY, PHYSICS

Pergamon Press
Maxwell House, Fairview Park
Elmsford, NY 10523

Editor: Dr. Philip Rubin, M.D., Department of Radiation, Oncology, Strong Memorial Hospital, 601 Elmwood Avenue, Box 647, Rochester, NY 14642
Phone: 716-275-5175
ISSN: 0360-3016 **Circulation:** 5,000 **Frequency:** M
Style Sheet: Yes
Revised Theses or Diss.: No **Manuscript Copies:** 5
Review Period: 3 months (PR) **Lagtime:** 6-12 months
Types of Articles: Clinical and basic research, reviews (usually solicited), technical innovations, notes, editorials (usually solicited), and special features (in conjunction with sponsoring societies).
Major Contents: Cancer research in the clinical and basic sciences involving: radiation, radiation combined with chemotherapy and/or surgery, physics and hyperthermia relating to cancer research and radiation oncology, clinical radiation oncology and protocol studies, radiation biology, hyperthermia, and radiation physics.
Miscellaneous: An official journal of the American Society for Therapeutic Radiology and Oncology.

276 INTERNATIONAL JOURNAL OF ROCK MECHANICS AND MINING SCIENCES AND GEOMECHANICS ABSTRACTS

Pergamon Press, Maxwell House
Fairview Park
Elmsford, NY 10523

Editor: Dr. E.T. Brown
Phone: 914-592-7700
ISSN: 0148-9062 **Circulation:** 1,400 **Frequency:** BM
Style Sheet: Yes
Revised Theses or Diss.: N/A **Manuscript Copies:** 3
Review Period: 1-4 months (PR) **Lagtime:** 12 months
Types of Articles: Technical notes, articles and book reviews.
Major Contents: Rock mechanics, mining sciences, and geomechanics.
Miscellaneous: Provides a medium of communication between the many individuals and groups of diverse background working in the general field of rock mechanics.

277 INTERNATIONAL JOURNAL OF SOLAR ENERGY

Harwood Academic Publishers
50 W. 23rd Street
New York, NY 10010

Editor: W. Palz, EEC Commission, Rue de lor Loi, B104G Brussels, Belgium
Phone: 212-242-4464
ISSN: 0142-5919 **Circulation:** N/A **Frequency:** BM
Style Sheet: Yes
Revised Theses or Diss.: Yes **Manuscript Copies:** 2
Review Period: 1-2 months (PR) **Lagtime:** 4-8 months
Types of Articles: Research, reviews, short notes, and projects.
Major Contents: Solar energy, biomass, wind energy, photovoltaics, solar heating, biofuels, and wind.

278 **INTERNATIONAL JOURNAL OF SOLIDS AND STRUCTURES**
Pergamon Press
Maxwell House, Fairview Park
Elmsford, NY 10523

Editor: George Herrmann
Phone: 914-592-7700
ISSN: 0020-7683 **Circulation:** 1,400 **Frequency:** M
Style Sheet: Yes
Revised Theses or Diss.: N/A **Manuscript Copies:** 2
Review Period: 2-3 months **Lagtime:** 6-12 months
Types of Articles: Manuscripts, reviews, brief articles, and discussions of previously published papers.
Major Contents: Mechanics of solids and structures including: fracture, flow, wave propagation, heat transfer, thermal effects in solids, optimum design methods and model analysis, structural topology, and numerical techniques.

279 **INTERNATIONAL JOURNAL OF SYSTEMATIC BACTERIOLOGY**
ASM Publications Department
1913 I Street, N.W.
Washington, DC 20006

Editor: W.E.C. Moore, Department of Anaerobic Microbiology, Virginia Polytechnic Institute, and State University, Blacksburg, VA 24061
Phone: 202-833-9680
ISSN: 0020-7713 **Circulation:** 2,000 **Frequency:** Q
Style Sheet: Council of Biology Editors.
Revised Theses or Diss.: Yes **Manuscript Copies:** 2
Review Period: 2-3 months (PR) **Lagtime:** N/A
Types of Articles: Original research not previously published.

Major Contents: Systematics of bacteria, yeasts, and yeastlike organisms, including taxonomy, nomenclature, identification, characterization, and culture preservation.
Miscellaneous: An official publication of the International Committee on Systematic Bacteriology of the International Union of Microbiological Societies. Manuscripts should be submitted directly to the ASM Publications Department (above).

280 **INTERNATIONAL JOURNAL OF TURBO AND JET-ENGINES**

Martinus Nijhoff Publishers
P.O. Box 163
3300 AD Dordrecht
The Netherlands

Editor: Professor B. Gal-Or, Editor-in-Chief, Head Turbo and Jet Engine Laboratory, Department of Aeronautical Engineering, Technion-IIT, Haifa, Israel
Phone: 04-292-807
ISSN: 0334-0082 **Circulation:** N/A **Frequency:** Q
Style Sheet: Yes
Revised Theses or Diss.: No **Manuscript Copies:** 3
Review Period: 2 months (PR) **Lagtime:** 6 months
Types of Articles: Research, reviews, and application papers.
Major Contents: Science and technology of air, land and marine applications of turbo and jet-engines including: fluid dynamics in turbomachinery, materials and modern processing methods, engine control, aero-thermodynamics, combustion, inlets, and compressors.
Miscellaneous: American Editor: Jean Louis, Department of Aeronautics and Astronautics, MIT, Cambridge, MA. European Editor: Roy Cookson, Cranfield Institute of Technology, Cranfield, Bedford, United Kingdom.

281 **INTERNATIONAL WATER POWER AND DAM CONSTRUCTION**

Business Press International, Ltd.
The Quadrant, Quadrant House
Sutton, Surrey
SM2 5AS, England

Editor: Falsian Acker
Phone: Surry 01-661-8815
ISSN: 0306-400X **Circulation:** 3,929 **Frequency:** M
Style Sheet: Yes
Revised Theses or Diss.: No **Manuscript Copies:** 1

Review Period: 3 months (PR) **Lagtime:** 6 months
Types of Articles: Descriptive, theoretical research, and development.
Major Contents: Dam construction, instrumentation, electrical system operation, evolution of water turbines and operation, control systems, and CAD related to dam construction.

282 IRISH JOURNAL OF EARTH SCIENCES

The Secretary, Royal Irish Academy
19 Dawson Street
Dublin 2, Ireland

ISSN: N/A **Circulation:** N/A **Frequency:** N/A
Style Sheet: Yes
Revised Theses or Diss.: No **Manuscript Copies:** 2
Review Period: 3-6 months (PR) **Lagtime:** 9-12 months
Types of Articles: Research and reviews.
Major Contents: Geological sciences.

283 JPT - JOURNAL OF PETROLEUM TECHNOLOGY

Society of Petroleum Engineers
P.O. Box 833836
Richardson, TX 75083

Editor: Jim McInnis
Phone: 214-669-3377
ISSN: 0149-2136 **Circulation:** 55,000 **Frequency:** M
Style Sheet: Society of Petroleum Engineers.
Revised Theses or Diss.: No **Manuscript Copies:** 3
Review Period: 3-6 months **Lagtime:** 3-9 months
Types of Articles: Projects, reviews, and field reports.
Major Contents: Drilling, exploration, production technology and economics. Professional engineering topics such as drilling-bits, telemetry, well control, casing, exploration-reservoir definition, reserves, production-completions, fracturing, well testing, and geothermal.
Miscellaneous: An official publication of the Society of Petroleum Engineers. Formerly: JOURNAL OF PETROLEUM TECHNOLOGY, ISSN: 0022-3522.

284 JOURNAL OF ADHESION, THE

Gordon and Breach Science Publishers, Inc.
1 Park Avenue
New York, NY 10010

Editor: Louis H. Sharpe

Phone: 201-582-4729
ISSN: 0021-8464 **Circulation:** N/A **Frequency:** Q
Style Sheet: Yes
Revised Theses or Diss.: No **Manuscript Copies:** 3
Review Period: 2-3 months (PR) **Lagtime:** 6 months
Types of Articles: Experimental, theoretical, and reviews.
Major Contents: Adhesion and the applications of adhesives.

285 JOURNAL OF ADVANCED TRANSPORTATION
The Institute of Transportation
1410 Duke University Road
Durham, NC 27701

Editor: Charles M. Harman, Department of Mechanical Engineering, Duke University, Durham, NC 27706
Phone: 919-549-0541
ISSN: 0197-6729 **Circulation:** 1,000 **Frequency:** 3/year
Style Sheet: Yes
Revised Theses or Diss.: No **Manuscript Copies:** 2
Review Period: 3 weeks (PR) **Lagtime:** 5 months
Types of Articles: Research, technical notes, and letters to the editors.
Major Contents: Topics on high-speed ground transport from such disciplines as economics, sociology, geography, environmental sciences, government, technology, and urban transportation.
Miscellaneous: Formerly: HIGH SPEED GROUND TRANSPORTATION JOURNAL.

286 JOURNAL OF AEROSOL SCIENCE
Pergamon Press
Maxwell House, Fairview Park
Elmsford, NY 10523

Editor: Dr. Dieter Hochrainer, Fraunhofer - Institut, ITA Grafschaft, D - 5948, Schmallenberg, Germany GFR
Phone: 914-592-7700
ISSN: 0021-8502 **Circulation:** 1,000 **Frequency:** BM
Style Sheet: Yes
Revised Theses or Diss.: Yes **Manuscript Copies:** 2
Review Period: N/A (PR) **Lagtime:** 6 months
Types of Articles: Research and reviews.
Major Contents: Scientific knowledge underlying applied problems in aerosol research including airborne infection, treatment of disease, exposure to dust, safety of nuclear reactors, radiation safety, climatic control, and chemical warfare.

287 JOURNAL OF AFRICAN EARTH SCIENCES
Pergamon Press
Maxwell House, Fairview Park
Elmsford, NY 10523

Editor: Professor C.A. Kogbe, Rock View International, CEEI - Rock View, 11 rus Jean-Mermoz, 75008 Paris, France
Phone: 914-592-7700
ISSN: 0731-7247 **Circulation:** N/A **Frequency:** Q
Style Sheet: Yes
Revised Theses or Diss.: Yes **Manuscript Copies:** 3
Review Period: N/A (PR) **Lagtime:** 6 months
Types of Articles: Book reviews, short notes, selected papers from proceedings of conferences, symposia, workshops, and software surveys.
Major Contents: All aspects of research related to the African continent including research on regional geology of African cratons, mobile belts, economic geology, geophysics, structure quarternary geology, and analysis of sedimentary basins.
Miscellaneous: Manuscripts may be submitted in English or French.

288 JOURNAL OF AGRICULTURAL AND FOOD CHEMISTRY
American Chemical Society
1155 16th Street, N.W.
Washington, DC 20036

Editor: Irvin E. Liener
Phone: 202-872-4542
ISSN: 0021-8561 **Circulation:** 4,800 **Frequency:** BM
Style Sheet: First issue of each volume.
Revised Theses or Diss.: Yes **Manuscript Copies:** 3
Review Period: 3-5 months (PR) **Lagtime:** 2-3 months
Types of Articles: Research, projects, reviews in multidisciplinary areas, and correspondence.
Major Contents: The chemistry of all aspects of agriculture and food including: pesticide chemistry, plant chemistry, plant nutrients and regulators, chemistry of food and feed, and biochemistry of nutrition.
Miscellaneous: Page charges of $40 per page may be paid, though payment does not affect acceptance.

289 JOURNAL OF AGRICULTURAL ENGINEERING
College of Agricultural Engineering
Punjab Agricultural University
Ludhiana-141004
Punjab, India

Editor: A.P. Bhatnagar, Chief Editor
Phone: India 22960-Ext. 384
ISSN: N/A **Circulation:** 1,500 **Frequency:** Q
Style Sheet: Yes
Revised Theses or Diss.: No **Manuscript Copies:** 3
Review Period: 6 months **Lagtime:** 3 months
Types of Articles: Research and development notes.
Major Contents: Farm machinery and power, soil and water engineering, energy-agriculture, processing, agricultural structures, and waste utilization.

290 **JOURNAL OF ANIMAL ECOLOGY, THE**
The British Ecological Society
Blackwell Scientific Publications, Ltd.
Osney Mead, Oxford
OX2 OEL, England

Editor: L.R. Taylor
ISSN: 0022-0477 **Circulation:** 3,970 **Frequency:** 3/year
Style Sheet: Yes
Revised Theses or Diss.: N/A **Manuscript Copies:** 2
Review Period: N/A **Lagtime:** N/A
Types of Articles: Research, projects, and reviews.
Major Contents: Research in the subject of ecology as a branch of natural science.

291 **JOURNAL OF ANIMAL SCIENCE**
American Society of Animal Science
309 W. Clark Street
Champaign, IL 61820

Editor: Dr. C.M. Bailey, Editor-in-Chief, Department of Animal Science, University of Nevada, Reno, Reno, NV 89557-0004
Phone: 702-784-1633
ISSN: 0021-8812 **Circulation:** 7,500 **Frequency:** M
Style Sheet: See Miscellaneous
Revised Theses or Diss.: Yes **Manuscript Copies:** 2
Review Period: 3 weeks (PR) **Lagtime:** 4-5 months
Types of Articles: Research and reviews.
Major Contents: All aspects of animal science or livestock production including breeding and genetics, nutrition (animal), physiology and endocrinology, growth and developmental biology, animal behavior, pharmacology toxicology, and meat science.

Miscellaneous: Council of Biological Editors Style Manual and Journal of Animal Science Style and Form.

292 **JOURNAL OF APPLIED ECOLOGY, THE**

The British Ecological Society
Blackwell Scientific Publications, Ltd.
Osney Mead, Oxford
OX2 OEL, England

Editor: Dr. T.M. Roberts, C.E.R.L. Biology Section, Kelvin Avenue, Leatherhead, Surrey KT22 7SE, England
Phone: Leatherhead 374488
ISSN: 0021-8901 **Circulation:** 2,000 **Frequency:** 3/year
Style Sheet: A Guide for Contributors.
Revised Theses or Diss.: No **Manuscript Copies:** 2
Review Period: 2-3 months (PR) **Lagtime:** 5-6 months
Types of Articles: Research, books, and reviews.
Major Contents: Research in the subject of ecology as a branch of natural science.

293 **JOURNAL OF APPLIED PHYSICS**

American Institute of Physics
335 E. 45th Street
New York, NY 10017

Editor: Dr. Gilbert J. Perlow, Journal of Applied Physics, Argonne National Laboratory, P.O. Box 296, Argonne, IL 60439
Phone: 212-661-9404
ISSN: 0021-8979 **Circulation:** 5,312 **Frequency:** M
Style Sheet: AIP Style Manual.
Revised Theses or Diss.: Yes **Manuscript Copies:** 2
Review Period: 1 month **Lagtime:** 4 months
Types of Articles: Research.
Major Contents: Applied physics.
Miscellaneous: Manuscripts should normally be limited to 1800 words, including an abstract.

294 **JOURNAL OF ARCHITECTURAL EDUCATION**

Assocication of Collegiate Schools of Architecture
1735 New York Avenue, N.W.
Washington, DC 20006

Editor: David H. Bell, Journal of Architechtural Education, School of Architecture, Rensselaer Polytechnic Institute, Troy, New York 12180

Phone: 202-785-2324
ISSN: 0047-2239 **Circulation:** 5,000 **Frequency:** 4/year
Style Sheet: Yes
Revised Theses or Diss.: Yes **Manuscript Copies:** 3
Review Period: Varies (PR) **Lagtime:** 6 months
Types of Articles: Research and articles.
Major Contents: Architecture and the built environment-some emphasis on educational aspects for the design professions. fessions. Energy issues, construction, history of architecture, design and theory, technology, teaching, professional practice, problem solving, and visual communication.

295 JOURNAL OF ATMOSPHERIC AND TERRESTRIAL PHYSICS

Pergamon Press
Maxwell House, Fairview Park
Elmsford, NY 10523

Editor: Professor Sir Granville Beynon FRS, Department of Physics, University College of Wales, Penglaris, Aberystwyth, Cardiganshire, England
Phone: 914-592-7700
ISSN: 0021-9169 **Circulation:** 1,350 **Frequency:** M
Style Sheet: Council of Biology Editors.
Revised Theses or Diss.: Yes **Manuscript Copies:** 2
Review Period: N/A (PR) **Lagtime:** 6 months
Types of Articles: Reviews and short papers.
Major Contents: Thunderstorm electricity and the electric field in the low atmosphere near the earth; problems of the magnetosphere, the ionosphere and aeronomy in the high atmosphere; aurora, geomagnetism, airflow, and radiowave propagation; electron density distributions and variations; electromagnetic radiation; and gravity waves.
Miscellaneous: Papers should be submitted to one of the following authors. B.H. Briggs, Department of Physics, University of Adelaide, S. Australia; B. Landmark, Royal Norwegian Council for Scientific & Industrial Research, Space Activity Division, Wdm. Thranes gt. 98, Oslo 1, Norway. J.S. Nisbet, Department of Electrical Engineering, Pennsylvania State University, University Park, PA 16802.

296 JOURNAL OF BACTERIOLOGY

American Society for Microbiology
1913 I Street, N.W.
Washington, DC 20006

Editor: Simon Silver
Phone: 202-833-9680
ISSN: 0021-9193 **Circulation:** 8,245 **Frequency:** M
Style Sheet: Council of Biology Editors.
Revised Theses or Diss.: Yes **Manuscript Copies:** 2
Review Period: 3 months (PR) **Lagtime:** 2-3 months
Types of Articles: Research.
Major Contents: Morphology and ultrastructure, genetics, molecular biology, physiology, metabolism, enzymology of bacteria, and other microorganisms.

297 JOURNAL OF BIOMECHANICS
Pergamon Press
Maxwell House, Fairview Park
Elmsford, NY 10523

Editor: Verne L. Roberts
Phone: 914-592-7700
ISSN: 0021-9290 **Circulation:** 1,600 **Frequency:** M
Style Sheet: Yes
Revised Theses or Diss.: N/A **Manuscript Copies:** 2
Review Period: 1-2 months (PR) **Lagtime:** 12 months
Types of Articles: Surveys, original papers, technical notes, and letters to the editors.
Major Contents: Human and non-human biomechanics from the medical, dental, basic sciences, and engineering fields.

298 JOURNAL OF CANADIAN PETROLEUM TECHNOLOGY
Canadian Institute of Mining and Metallurgy
1130 Sherbrooke Street, W.
Suite 400
Montreal, Quebec H3A 2M8, Canada

Editor: A. Kerr and P. Michaud
Phone: 514-842-3461
ISSN: 0021-9487 **Circulation:** 9,000 **Frequency:** BM
Style Sheet: Yes
Revised Theses or Diss.: Yes **Manuscript Copies:** 3
Review Period: 6-18 months (PR) **Lagtime:** 6-18 months
Types of Articles: Research, projects, and reviews.
Major Contents: Technical information regarding all aspects of exploration, drilling, production and reservoir engineering in the petroleum and natural gas industry.

Miscellaneous: Published by the Canadian Institute of Mining and Metallurgy for the Petroleum Society of CIM.

299 JOURNAL OF CHEMICAL EDUCATION
American Chemical Society
1155 16th Street, N.W.
Washington, DC 20036

Editor: J.J. Lagowski, University of Texas, Department of Chemistry, Austin, TX 78712
Phone: 512-471-3288
ISSN: 0021-9584 **Circulation:** 21,000 **Frequency:** M
Style Sheet: Yes
Revised Theses or Diss.: Yes **Manuscript Copies:** 3
Review Period: 1-2 months (PR) **Lagtime:** 12-15 months
Types of Articles: Research, projects, reviews, and descriptions of lab experiments and procedures.
Major Contents: Reviews of all areas of chemistry with a slant toward classroom applications, and new curricula. features on safety, apparatus, and teaching techniques.

300 JOURNAL OF CLIMATE AND APPLIED METEOROLOGY
American Meteorological Society
45 Beacon Street
Boston, MA 02108

Editor: Bernard Silverman and Alan Hecht
Phone: 617-227-2425
ISSN: 0733-3021 **Circulation:** 4,100 **Frequency:** M
Style Sheet: Yes
Revised Theses or Diss.: N/A **Manuscript Copies:** 5
Review Period: 1-2 months (PR) **Lagtime:** 7-8 months
Types of Articles: Articles, notes, and correspondence.
Major Contents: Application of atmospheric sciences to the health, economy, and well-being of the human community. Transportation, communication, environmental health, air pollution, meteorology, weather modification, agricultural and forest meteorology, and forecasting.

301 JOURNAL OF COMPOSITE MATERIALS
Technomic Publishing Company, Inc.
265 Post Road, W.
Westport, CT 06880

Editor: Dr. Stephen Tsai

Phone: 717-291-5609
ISSN: 0021-9983 **Circulation:** 1,200 **Frequency:** BM
Style Sheet: Yes
Revised Theses or Diss.: N/A **Manuscript Copies:** 2
Review Period: 2 months (PR) **Lagtime:** 6 months
Types of Articles: Research.
Major Contents: Theoretical and experimental studies of the physical and structural properties of multiphase materials, with emphasis on phenomenological and mechanistic approaches to: fracture, fatigue, structural reliability, and design criteria.

302 JOURNAL OF CRYSTAL GROWTH

Elsevier Science Publishers B.V.
P.O. Box 211
1000 AE Amsterdam
The Netherlands

Editor: Editorial Board
ISSN: 0022-0248 **Circulation:** N/A **Frequency:** 18/year
Style Sheet: Yes
Revised Theses or Diss.: N/A **Manuscript Copies:** 3
Review Period: 2 weeks **Lagtime:** 6 months
Types of Articles: Full-length papers, conference proceedings, book reviews, news, announcements, and letters to the editors.
Major Contents: Nucleation and growth, molecular kinetics and transport phenomena, crystallization in viscous media, crystal growth of metals, semiconductors, magnetics, inorganic, organic, and biological substances, techniques for crystal growth, and characterization of single crystals.

303 JOURNAL OF ELECTROCHEMICAL SOCIETY

Electrochemical Society, Inc.
10 S. Main Street
Pennington, NJ 08534

Editor: Sarah A. Kilfoyle, Director of Publications
Phone: 609-737-1902
ISSN: 0013-4651 **Circulation:** 8,200 **Frequency:** M
Style Sheet: Yes
Revised Theses or Diss.: Yes **Manuscript Copies:** 3
Review Period: 3 months (PR) **Lagtime:** 10 months
Types of Articles: Technical papers, technical notes, brief communications, and reviews.

Major Contents: Electrochemical science and technology, solid state science and technology.

304 **JOURNAL OF ELECTROSTATICS**
Elsevier Science Publishers B.V.
P.O. Box 330
1000 AH Amsterdam
The Netherlands

Editor: D.K. Davies, Applied Physics Department, ERA Technology Ltd., Leatherhead, Cleeve Road, Surrey, KT22 7SA Great Britain, The Netherlands
Phone: 020-5803-911
ISSN: 0304-3886 **Circulation:** N/A **Frequency:** BM
Style Sheet: Yes
Revised Theses or Diss.: No **Manuscript Copies:** 3
Review Period: N/A (PR) **Lagtime:** 6-12 months
Types of Articles: Reviews, communications, and research.
Major Contents: Topics on static electricity in its fundamental aspects, hazardous nature, and applications, such as: field problems, electrostatic effects in solids, electrostatic machines, effects in biological materials and in gases such as plasmas, and instrumentation for measurement of static-electricity phenomena.

305 **JOURNAL OF ENERGY AND DEVELOPMENT**
University of Colorado
216 Economics Building, Box 263
Boulder, CO 80309

Editor: Ragaei El Mallakh
Phone: 303-492-7667
ISSN: 0361-4476 **Circulation:** 2,000 **Frequency:** SA
Style Sheet: Chicago Manual of Style.
Revised Theses or Diss.: Yes **Manuscript Copies:** 2
Review Period: 4-6 months (PR) **Lagtime:** 4-12 months
Types of Articles: Research, projects, and occasional reviews.
Major Contents: Economics of energy development, conservation, environment, self-sufficiency, interdependence, resources ownership and control policies, fiscal and monetary aspects of energy, economic planning in energy, growth and stability in energy-producing and importing countries, alternate energy sources to petroleum, and multinational energy entities.
Miscellaneous: The Journal addresses itself to professionals, members of the business community and industry, government officials, and

scholars in economics and related areas concerned with energy.

306 **JOURNAL OF ENERGY ENGINEERING**
American Society of Civil Engineers
345 E. 47th Street
New York, NY 10017

Editor: Edgar T. Moore, Jr.
Phone: 212-705-7496
ISSN: 0733-9402 **Circulation:** N/A **Frequency:** IR
Style Sheet: *See Miscellaneous
Revised Theses or Diss.: Yes **Manuscript Copies:** 3
Review Period: 3 months (PR) **Lagtime:** 15 months
Types of Articles: Research, projects, and reviews.
Major Contents: Focuses on the development of scientific and engineering knowledge in the planning, management, and generation of electric power, the development of nonconventional energy sources and the environmental effects and means for conserving natural resources. Emphasis is on fossil power, nuclear power, radioactive waste disposal, hydropower planning and design.
Miscellaneous: Formerly: JOURNAL OF THE ENERGY DIVISION. An official publication of the American Society of Civil Engineers, Energy Division. *Author's Guide to Publications of the ASCE.

307 **JOURNAL OF ENERGY LAW AND POLICY**
University of Utah
College of Law
Salt Lake City, UT 84112

Editor: Robert L. Tamietti, Articles Editor
Phone: 801-581-6833
ISSN: 0275-9926 **Circulation:** 1,000 **Frequency:** SA
Style Sheet: A Uniform System of Citation.
Revised Theses or Diss.: Yes **Manuscript Copies:** 1
Review Period: 1 month (PR) **Lagtime:** 4-6 months
Types of Articles: Research, projects, and book reviews.
Major Contents: Energy, natural resources and related legal topics of interest to lawyers and energy policy makers.

308 **JOURNAL OF ENERGY RESOURCES TECHNOLOGY**
American Society of Mechanical Engineers
345 E. 47th Street
New York, NY 10017

Editor: Jin S. Chung, Technical Editor, Department of Engineering, Colorado School of Mines, 1500 Illinois Street, Golden, CO 80401
Phone: 303-273-3673
ISSN: 0195-0738 **Circulation:** 1,900 **Frequency:** Q
Style Sheet: An ASME Paper, ASME Manual MS-4.
Revised Theses or Diss.: Yes **Manuscript Copies:** 4
Review Period: 2-3 months **Lagtime:** 6 months
Types of Articles: Research and engineering reports, reviews for engineers and designers in the energy resources, offshore mechanics, and Arctic engineering fields.
Major Contents: Developing energy resources industry technologies: e.g. renewable energy, rock mechanics, drilling mechanics, petroleum fluid mechanics, offshore mechanics/control, offshore technology, computer technology, Arctic engineering, ice mechanics, and energy from oceans. Includes research, engineering and design data, and recent advances in frontier technology.
Miscellaneous: An official publication of the American Society of Mechanical Engineers.

309 JOURNAL OF ENVIRONMENTAL SCIENCE AND HEALTH: PART A ENVIRONMENTAL SCIENCE AND ENGINEERING

Marcel Dekker, Inc.
270 Madison Avenue
New York, NY 10016

Editor: Dr. James W. Robinson, Chemistry Department, Louisiana State University, Baton Rouge, LA 70803
Phone: 504-388-3025
ISSN: 0360-1226 **Circulation:** N/A **Frequency:** 8/year
Style Sheet: Yes
Revised Theses or Diss.: No **Manuscript Copies:** 2
Review Period: 3-6 weeks (PR) **Lagtime:** 3 months
Types of Articles: Research and reviews.
Major Contents: All aspects related to the environment, including legal and medical. Particular attention is given to studies involving toxic and hazardous substances disposal and control.
Miscellaneous: An international journal. No advertising.

310 JOURNAL OF ENVIRONMENTAL SCIENCE AND HEALTH: PART B PESTICIDES, FOOD CONTAMINANTS, AND AGRICULTURAL WASTES

Marcel Dekker, Inc.
270 Madison Avenue
New York, NY 10016

Editor: Shahamat U. Khan, Chemistry and Biology Research Institute, Research Branch, Canada Agriculture, Ottawa, Ontario, K1A OC6, Canada
ISSN: 0360-1234 **Circulation:** N/A **Frequency:** 9/year
Style Sheet: Yes
Revised Theses or Diss.: N/A **Manuscript Copies:** 2
Review Period: N/A **Lagtime:** N/A
Types of Articles: Original research and reports.
Major Contents: Topics concerning analytical techniques or improvements on existing methods applicable to residues of pesticides, food contaminants (natural and additive), and other chemical contaminants, and their metabolites in the ecosphere; original factual reports on the persistence, binding, translocation, chemical and biodegradation, methods of detoxification; and their toxicological considerations and consequences. The journal will also encompass the developments in integrated methods of pest control and various aspects of agricultural wastes.

311 JOURNAL OF ENVIRONMENTAL SCIENCE AND HEALTH: PART C ENVIRONMENTAL CARCINOGENESIS REVIEWS
Marcel Dekker, Inc.
270 Madison Avenue
New York, NY 10016

Editor: Joseph C. Arcos, Mary F. Argus, & Yin-Tak Woo, P.O. Box 2429, Springfield, VA 22152
ISSN: 0736-3001 **Circulation:** N/A **Frequency:** 2/year
Style Sheet: Yes
Revised Theses or Diss.: N/A **Manuscript Copies:** 2
Review Period: N/A (PR) **Lagtime:** N/A
Types of Articles: Research.
Major Contents: Topics include reviews on specific compounds or classes of compounds of special interest; structure-activity correlation analysis; theoretical models; synergism and antagonism; promotion and cocarcinogenesis; interaction of physical, chemical and biological factors; modification of carcinogenesis; metabolism and mechanism of action; source, occurrence, formation and exposure in the environment and in the workplace; risk assessment; epidemiological studies; correlation of carcinogenesis and mutagenesis.
Miscellaneous: Articles are solicited by invitation and the editors welcome suggestions about prospective authors and topics. Contributors interested in submitting articles on their own initiative are encouraged to submit a letter of inquiry giving the title and a brief outline of the

proposed review so the Editorial Board can determine whether the topic is appropriate and unassigned to other reviewers.

312 JOURNAL OF ENVIRONMENTAL SCIENCES, THE
Institute of Environmental Sciences
940 E. Northwest Highway
Mt. Prospect, IL 60056

Editor: Henry Caruso
Phone: 312-255-1561
ISSN: 0022-0906 **Circulation:** 3,000 **Frequency:** BM
Style Sheet: Yes
Revised Theses or Diss.: Yes **Manuscript Copies:** 3
Review Period: 6 months **Lagtime:** 6-12 months
Types of Articles: Research and projects.
Major Contents: Topics dealing with test design and evaluation, products and reliability, and energy and the environment.

313 JOURNAL OF ENVIRONMENTAL SYSTEMS
Baywood Publishing Company, Inc.
120 Marine Street, Box D
Farmingdale, NY 11735

Editor: Paul R. DeCicco
Phone: 516-249-2464
ISSN: 0047-2433 **Circulation:** N/A **Frequency:** Q
Style Sheet: IFT Style Guide for Research Papers
Revised Theses or Diss.: N/A **Manuscript Copies:** 3
Review Period: 1 month (PR) **Lagtime:** 5 months
Types of Articles: Research and book reviews.
Major Contents: Analysis, design, and management of the environment.

314 JOURNAL OF FOOD SCIENCE
Institute of Food Technologists
221 N. LaSalle Street
Chicago, IL 60601

Editor: Dr. Aaron E. Wasserman, IFT Scientific Editor, P.O. Box 6197, Philadelphia, PA 19115
Phone: 312-782-8424
ISSN: 0022-1147 **Circulation:** 10,000 **Frequency:** BM
Style Sheet: IFT Guide for Research Papers.
Revised Theses or Diss.: N/A **Manuscript Copies:** 3

Review Period: N/A **Lagtime:** N/A
Types of Articles: Research and reviews.
Major Contents: Applied science and engineering and basic research concerning food sciences.

315 JOURNAL OF FOOD TECHNOLOGY

Blackwell Scientific Publications, Ltd.
Osney Mead, Oxford OX2 OEL
England

Editor: H. Liebmann, Institute of Food Science and Technology, 20 Queensberry Place, London SW7 2DR, England
ISSN: 0022-1163 **Circulation:** 2,900 **Frequency:** BM
Style Sheet: Yes
Revised Theses or Diss.: Yes **Manuscript Copies:** 3
Review Period: 2-3 months **Lagtime:** 5 months
Types of Articles: Research, projects, reviews, literature reviews, and technical notes.
Major Contents: Physical, chemical, microbiological and nutritional aspects of foods, food processing and preservation, engineering aspects of food processing and preservation.
Miscellaneous: Publication of the Institute of Food Science and Technology (U.K.).

316 JOURNAL OF FUSION ENERGY

Plenum Publishing Corporation
233 Spring Street
New York, NY 10013

Editor: Lawrence M. Lidsky, Plasma Fusion Center, Room 38-174, Massachusetts Institute of Technology, Cambridge, MA 02139
Phone: 212-741-6680
ISSN: 0164-0313 **Circulation:** N/A **Frequency:** BM
Style Sheet: Yes
Revised Theses or Diss.: Yes **Manuscript Copies:** 3
Review Period: 1-2 months (PR) **Lagtime:** 6-12 months
Types of Articles: Research and reviews.
Major Contents: Thermonuclear fusion as a power source, research results in thermonuclear fusion, and policy and planning in the fusion program.

317 JOURNAL OF GEOLOGICAL EDUCATION

National Association of Geology Teachers
Allen Press, Inc.
Box 368
Lawrence, KS 66044

Editor: J.H. Shea, University of Wisconsin, Parkside, P.O. Box 2000, Kenosha, WI 53141
Phone: 913-843-1234
ISSN: 0022-1368 **Circulation:** 2,600 **Frequency:** 5/year
Style Sheet: No
Revised Theses or Diss.: No **Manuscript Copies:** 1
Review Period: 6 months **Lagtime:** 6-8 months
Types of Articles: Research, projects, and reviews.
Major Contents: Geology, minerals, paleontology, earth science education, history of geology, geomorphology, mineral deposits, and economic geology.

318 **JOURNAL OF GEOLOGY**
University of Chicago Press
5801 Ellis Avenue
Chicago, IL 60637

Editor: Alfred T. Anderson and Robert C. Newton
Phone: 312-962-8138
ISSN: 0022-1376 **Circulation:** 3,400 **Frequency:** BM
Style Sheet: Yes
Revised Theses or Diss.: Yes **Manuscript Copies:** 3
Review Period: 2 months (PR) **Lagtime:** 9 months
Types of Articles: Research and geological notes.
Major Contents: All aspects of geology of general as opposed to local or regional interest including geophysics, geochemistry, structural geology, mineralogy/ petrology, paleontology, oceanography, paleoclimatology, and geomorphology.
Miscellaneous: Theses and dissertations accepted only if in regular research. Small contributions (10 printed pages or less) of slightly more specialized nature may be accepted for the geological notes section.

319 **JOURNAL OF GEOPHYSICAL RESEARCH**
Department of Earth and Space Physics
University of California
Los Angeles, CA 90024

Editor: Gerald Schubert
Phone: 213-825-4577
ISSN: 0196-6928 **Circulation:** 4,900 **Frequency:** M
Style Sheet: AGU Style Guide.
Revised Theses or Diss.: Yes **Manuscript Copies:** 4

Review Period: 1-2 months (PR) **Lagtime:** 9 months
Types of Articles: Research, theoretical, and projects.
Major Contents: Physics and chemistry of the earth, its environments, and the solar system.
Miscellaneous: Information for contributors published in January issue.

320 JOURNAL OF HEAT RECOVERY SYSTEMS

Pergamon Press
Maxwell House, Fairview Park
Elmsford, NY 10523

Editor: Daniel Reay, Editor-in-Chief, Journal of Heat Recovery Systems, IRD Company, Ltd., Fossway, Newcastle upon Tyne NE6 2YD, England
Phone: 091-265-0451
ISSN: 0198-7593 **Circulation:** N/A **Frequency:** BM
Style Sheet: No
Revised Theses or Diss.: Yes **Manuscript Copies:** 2
Review Period: 1-2 months (PR) **Lagtime:** 3-5 months
Types of Articles: Projects, case histories, reviews, R&D, and developments.
Major Contents: Heat exchangers, heat recovery, and heat pumps.

321 JOURNAL OF INDUSTRIAL IRRADIATION TECHNOLOGY

Marcel Dekker, Inc.
270 Madison Avenue
New York, NY 10016

Editor: R. Bradley, Director, Research and Development, Squibb Convatec, 25 Cotterslane, East Brunswick, NJ 08816
Phone: 201-254-8786
ISSN: 0735-7923 **Circulation:** N/A **Frequency:** Q
Style Sheet: Yes
Revised Theses or Diss.: Yes **Manuscript Copies:** 3
Review Period: 3 months (PR) **Lagtime:** 6 months
Types of Articles: Research letters including review, development, and economics.
Major Contents: All facets of irradiation technology including: engineering, production, economics, marketing, research and development, applications of ionizing radiation to industrial processes. Emphasis is on the practical, and industrial uses of ionizing radiation.
Miscellaneous: Letters, short commentaries, as well as disscussions-in-print of controversial topics are encouraged.

322 JOURNAL OF LUMINESCENCE
Elsevier Science Publishers B.V.
P.O. Box 211
1000 AE Amsterdam
The Netherlands

Editor: F. Williams
Phone: 020-5803-911
ISSN: 0022-2313 **Circulation:** N/A **Frequency:** 8/year
Style Sheet: Yes
Revised Theses or Diss.: N/A **Manuscript Copies:** 2
Review Period: 1-2 months (PR) **Lagtime:** 5 months
Types of Articles: Research, reviews, short communications, letters to the editor, book reviews, and news announcements.
Major Contents: General topics in luminescence including: coherent and incoherent emission, all materials that luminesce, and luminescent measurement techniques.

323 JOURNAL OF MAGNETISM AND MAGNETIC MATERIALS
Elsevier Science Publishers B.V.
P.O. Box 211
1000 AE Amsterdam
The Netherlands

Editor: A.J. Freeman
Phone: 020-5803-911
ISSN: 0304-8853 **Circulation:** 800 **Frequency:** 18/year
Style Sheet: Yes
Revised Theses or Diss.: N/A **Manuscript Copies:** 2
Review Period: 2 weeks (PR) **Lagtime:** 6 months
Types of Articles: Research, reviews, short notes, and letters to the editor.
Major Contents: Basic magnetism, the technology and applications of magnetic materials in the study of magnetism, and magnetic materials.

324 JOURNAL OF MATERIALS FOR ENERGY SYSTEMS
American Society for Metals
Energy Division
Metals Park, OH 44073

Editor: Suzanne Walker, Managing Editor
Phone: 216-338-5151
ISSN: 0162-9719 **Circulation:** 1,100 **Frequency:** Q
Style Sheet: Yes

Revised Theses or Diss.: Yes **Manuscript Copies:** 3
Review Period: 3 months (PR) **Lagtime:** 3 months
Types of Articles: Research, projects, and reviews.
Major Contents: New energy technologies, production processes, conservation systems for fossil fuel, hydroelectric, fusion, fission, solar, geothermal, ocean thermal. Other regenerative energy technologies, and construction and related materials used in energy systems components.
Miscellaneous: This is not a journal about energy production, but of the materials used in the construction of systems which produce energy.

325 **JOURNAL OF MATERIALS SCIENCE**

Chapman and Hall, Ltd.
11 New Fetter Lane
London EC4 PEE
England

Editor: W. Bonfield
ISSN: 0022-2461 **Circulation:** N/A **Frequency:** M
Style Sheet: Yes
Revised Theses or Diss.: N/A **Manuscript Copies:** 2
Review Period: Varies **Lagtime:** Varies
Types of Articles: Research, short communications, discussions on previous papers, letters to the editor, reviews, and annotated bibliographies.
Major Contents: Topics on the relationships between the structure, properties, and uses of materials - such as composite materials research, research on materials on which unusual demands are made, and papers on conventional materials.

326 **JOURNAL OF METAMORPHIC GEOLOGY**

Blackwell Scientific Publications, Ltd.
Osney Mead, Oxford OX2 OEL
England

Editor: Professor T.P. Loomis, Department of Geoscience, The University of Arizona, Tucson, AZ 85721
Phone: 602-621-6007
ISSN: 0263-4929 **Circulation:** 220 **Frequency:** Q
Style Sheet: Yes
Revised Theses or Diss.: Yes **Manuscript Copies:** 3
Review Period: 2 months (PR) **Lagtime:** 3-6 months
Types of Articles: Research, projects, and reviews.

Major Contents: Entire scope of metamorphic studies from the scale of the individual crystal to the lithospheric plate-properties of metamorphic minerals, theoretical and experimental studies of metamorphic reactions, and geochemistry of metamorphic rocks.

327 **JOURNAL OF MICROWAVE POWER**

International Microwave Power Institute
301 Maple Avenue, W., Suite 520
Vienna, VA 22180

Editor: Satish Kashop
Phone: 703-281-1515
ISSN: 0022-2739 **Circulation:** 1,000 **Frequency:** Q
Style Sheet: Yes
Revised Theses or Diss.: N/A **Manuscript Copies:** 3
Review Period: 1-2 months (PR) **Lagtime:** 6-9 months
Types of Articles: Papers, brief communications or letters, reviews, technical analysis, and scientific research.
Major Contents: Topics in microwave energy and microwave energy applications.

328 **JOURNAL OF MINES, METALS AND FUELS**

6/2 Madan Street
Calcutta 700 072, India

Editor: A.K. Ghose
ISSN: 0022-2755 **Circulation:** 2,740 **Frequency:** M
Style Sheet: Yes
Revised Theses or Diss.: No **Manuscript Copies:** 3
Review Period: 1 month **Lagtime:** 2-3 months
Types of Articles: Research, projects, and reviews.
Major Contents: Mining, mineral beneficiation, metallurgy, bulk material handling, geology, coal and petrochemical, and fuel.

329 **JOURNAL OF NON-NEWTONIAN FLUID MECHANICS**

Elsevier Science Publishers B.V.
P.O. Box 211
1000 AE Amsterdam
The Netherlands

Editor: K. Walters
Phone: 020-5803-911
ISSN: 0377-0257 **Circulation:** N/A **Frequency:** 9/year
Style Sheet: Yes

Revised Theses or Diss.: N/A **Manuscript Copies:** 2
Review Period: 1-2 months (PR) **Lagtime:** 3-6 months
Types of Articles: Research.
Major Contents: Topics in basic rheological science such as: development of rheological equations of state for non-newtonian liquids, use of rheological equations of state to solve flow problems of one or more fluid phases, experiments on non-newtonian liquids, heat and mass transfers in non-Newtonian liquids, and instabilities in non-Newtonian fluids.

330 JOURNAL OF NUCLEAR MATERIALS

Elsevier Science Publishers B.V.
P.O. Box 211
1000 AE Amsterdam
The Netherlands

Phone: 020-5803-911
ISSN: 0022-3115 **Circulation:** N/A **Frequency:** SM
Style Sheet: N/A
Revised Theses or Diss.: N/A **Manuscript Copies:** N/A
Review Period: 2-4 weeks **Lagtime:** 4-6 months
Types of Articles: Reviews, research, letters to the editors, news, and announcements.
Major Contents: Materials for nuclear science; technology and irradiation effects in solids, nuclear fuels and fuel elements; structural and control materials; irradiation affects in fission and fusion reactor materials; and other metallic and non-metallic solids.

331 JOURNAL OF PETROLEUM GEOLOGY

The Scientific Press, Ltd.
P.O. Box 21
Beaconsfield Bucks HP9 1NS
England

Editor: Dr. E.N. Tiratsoo
Phone: 049-46-5139
ISSN: 0141-6421 **Circulation:** N/A **Frequency:** BM
Style Sheet: Yes
Revised Theses or Diss.: Yes **Manuscript Copies:** 2
Review Period: 1-3 months (PR) **Lagtime:** 3-6 months
Types of Articles: Research, projects, and reviews.
Major Contents: All aspects of the genesis and accumulation of petroleum and geology of oil and gas fields. Editorial preference is for papers on oil field regions outside North America and topics of general appli-

cation in petroleum exploration and development.

332 **JOURNAL OF PETROLOGY**
Oxford University Press
Walton Street
Oxford OX2 6DP
England

Editor: Dr. I.S. McCallum or Dr. B.W. Evans, Department of Geological Sciences, AJ-20, University of Washington, Seattle, WA 98195
Phone: 212-581-4300
ISSN: 0022-3530 **Circulation:** 1,600 **Frequency:** Q
Style Sheet: Yes
Revised Theses or Diss.: No **Manuscript Copies:** 3
Review Period: 1 month (PR) **Lagtime:** 12 months
Types of Articles: Research and reviews.
Major Contents: All aspects of mineralogy, geochemistry, geochronology, rock microstructure, and petrology; usually igneous and metamorphic.

333 **JOURNAL OF PHYSICS AND CHEMISTRY OF SOLIDS**
Pergamon Press
Maxwell House, Fairview Park
Elmsford, NY 10523

Editor: David E. Cox, Brookhaven National laboratory, Upton, NY 11973
Phone: 914-592-7700
ISSN: 0022-3697 **Circulation:** 2,300 **Frequency:** M
Style Sheet: See Miscellaneous
Revised Theses or Diss.: Yes **Manuscript Copies:** 2
Review Period: 2 months (PR) **Lagtime:** 6 months
Types of Articles: Research, short communications as technical notes, and invited topical reviews.
Major Contents: Covers all aspects of the fundamental physics and chemistry of the solid state including electronic, spectroscopic and structural properties of solids, and the statistical mechanics and thermodynamics of condensed systems.
Miscellaneous: Syle requirements are in each issue.

334 **JOURNAL OF POWER SOURCES**
Elsevier Sequoia S.A.
P.O. Box 851
CH-1001 Lausanne 1
Switzerland

Editor: D.H. Collins
ISSN: 0378-7753 **Circulation:** N/A **Frequency:** 8/year
Style Sheet: Yes
Revised Theses or Diss.: N/A **Manuscript Copies:** 3
Review Period: 2 months **Lagtime:** 5-6 months
Types of Articles: Research, communications, and reviews.
Major Contents: Batteries, fuel cells, solar power sources, thermionic, thermo-electric, isotope generators, other forms of nonmechanical generation, storage, and conversion of electricity.

335 JOURNAL OF PROPULSION AND POWER
American Institute of Aeronautics and Astronautics
1633 Broadway Street
New York, NY 10019

Editor: Professor Gordon C. Oates, Editor-in-Chief, Journal of Propulsion and Power, FS-10, University of Washington, Seattle, WA 98195
Phone: 212-581-4300
ISSN: 0748-4658 **Circulation:** 1,500 **Frequency:** BM
Style Sheet: Yes
Revised Theses or Diss.: Yes **Manuscript Copies:** 4
Review Period: 3 months (PR) **Lagtime:** 2-4 months
Types of Articles: Research, reviews, and applications.
Major Contents: This journal is devoted to the advancement of the science and technology of aerospace propulsion and power through the dissemination of original archival papers contributing to advancements in airbreathing, electric, and exotic propulsion; solid and liquid rockets, fuels and propellants, power generation and conversion for aerospace vehicles, and the application of aerospace science and technology to terrestrial energy devices and systems. It is intended to provide readers of the Journal, with primary interests in propulsion and power, access to papers spanning the range from research through development to applications. Papers in these disciplines and the related sciences of combustion, fluid mechanics, and solid mechanics as directly related to propulsion and power are solicited.
Miscellaneous: An official publication of the American Institute of Aeronautics and Astronautics.

336 JOURNAL OF PROTECTIVE COATINGS AND LININGS
Steel Structures Painting Council
4400 Fifth Avenue
Pittsburg, PA 15213

Editor: Harold E. Hower
Phone: 412-578-3106
ISSN: 8755-1985 **Circulation:** 10,000 **Frequency:** M
Revised Theses or Diss.: Yes **Manuscript Copies:** 1
Review Period: 2 months (PR) **Lagtime:** 2 months
Types of Articles: Research, problem solving forum, new technologies and applications, products and equipment descriptions and literature reviews.
Major Contents: All aspects used for corrosion control, protective coatings and linings, the materials from which they are made, their properties, and applications.
Miscellaneous: Articles do not necessarily reflect the opinions of the Steel Structures Painting council. Articles are generally more applied than research oriented.

337 **JOURNAL OF RADIOANALYTICAL AND NUCLEAR CHEMISTRY**
Elsevier Sequoia S.A.
P.O. Box 851
CH-1001 Lausanne 1
Switzerland

Editor: Professor T. Braun, Editor-in-Chief, Institute of Inorganic & Analytical Chemistry, Eotvos University, P.O. Box 123, 1443 Budapest, Hungary
Phone: 021-20-73-81
ISSN: 0236-5731 **Circulation:** N/A **Frequency:** 28/year
Style Sheet: Yes
Revised Theses or Diss.: No **Manuscript Copies:** 2-3
Review Period: 3-4 months (PR) **Lagtime:** 6-8 months
Types of Articles: Research, reviews, short communications, and letters.
Major Contents: Nuclear chemistry, radiochemistry, radiation chemistry, radiobiological chemistry, environmental radiochemistry, nuclear power plant chemistry, and radiation detection and measurement.
Miscellaneous: Formerly: JOURNAL OF RADIOANALYTICAL CHEMISTRY. Published under two separate covers, "Articles (8 + per year) and "Letters" (18 + per year).

338 **JOURNAL OF RHEOLOGY**
John Wiley and Sons, Inc.
605 Third Avenue
New York, NY 10016

Editor: Raymond R. Myers

Phone: 212-850-6000
ISSN: 0148-6055 **Circulation:** 1,679 **Frequency:** BM
Style Sheet: Yes
Revised Theses or Diss.: N/A **Manuscript Copies:** 3
Review Period: 6 months (PR) **Lagtime:** 12 months
Types of Articles: Research.
Major Contents: Topics on the mechanical properties of matter: primarily flow and deformation of matter.
Miscellaneous: Published for The Society of Rheology.

339 JOURNAL OF SOLAR ENERGY ENGINEERING
American Society of Mechanical Engineers
345 E. 47th Street
New York, NY 10017

Editor: Cornelia Monahan, Managing Editor
Phone: 212-705-7722
ISSN: 0199-6231 **Circulation:** 1,670 **Frequency:** Q
Style Sheet: ASME MS-4.
Revised Theses or Diss.: N/A **Manuscript Copies:** 5
Review Period: N/A (PR) **Lagtime:** N/A
Types of Articles: Technical papers and briefs.
Major Contents: All aspects of solar energy as well as selected aspects of wind and tidal energy. Areas of interest include, but are not limited to, biomass production and conversion, passive solar design, solar radiation, solar collectors and concentrators, photovoltaics, conversion processes, power systems, ocean energy systems, resource and availability studies, wind and tidal systems, and heat storage.
Miscellaneous: An official Journal of the American Society of Mechanical Engineers.

340 JOURNAL OF STRUCTURAL MECHANICS
Marcel Dekker, Inc.
270 Madison Avenue
New York, NY 10016

Editor: E. Hawg
Phone: 212-696-9000
ISSN: 0360-1218 **Circulation:** N/A **Frequency:** Q
Style Sheet: Yes
Revised Theses or Diss.: N/A **Manuscript Copies:** 3
Review Period: 1 month **Lagtime:** 12 months
Types of Articles: Research and reviews.

Major Contents: Structural mechanics which may be applicable to structural engineering.
Miscellaneous: Formerly: STRUCTURAL MECHANICS, ISSN: 0039-2561.

341 JOURNAL OF THE AIR POLLUTION CONTROL ASSOCIATION (JAPCA)

Air Pollution Control Association
P.O. Box 2861
Pittsburgh, PA 15230

Editor: Harold Englund
Phone: 412-621-1090
ISSN: 0002-2470 **Circulation:** 10,000 **Frequency:** M
Style Sheet: N/A
Revised Theses or Diss.: N/A **Manuscript Copies:** N/A
Review Period: N/A **Lagtime:** N/A
Types of Articles: Technical papers, research, and book reviews.
Major Contents: Air pollution and its control, and hazardous waste management.

342 JOURNAL OF THE AMERICAN OIL CHEMISTS SOCIETY (JAOCS)

American Oil Chemists' Society
508 S. 6th Street
Champaign, IL 61820

Editor: Dr. A.R. Baldwin, 4854 Thomas Avenue, S., Minneapolis, MN 55410
Phone: 217-359-2344
ISSN: 0003-021X **Circulation:** 7,000 **Frequency:** M
Style Sheet: Yes
Revised Theses or Diss.: Yes **Manuscript Copies:** 3
Review Period: 1-6 months (PR) **Lagtime:** 1-2 months
Types of Articles: Research, some reviews, communications, methodology, and comments to the editor on previous papers.
Major Contents: Anything pertaining to animal and vegetable fats and oils and their derivatives including analysis, processing methodology, utilization, and metabolism.
Miscellaneous: Formerly: OIL AND SOAP.

343 JOURNAL OF THE ASTRONAUTICAL SCIENCES, THE

American Astronautical Society, Inc.
6060 Tower Court
Alexandria, VA 22304

Editor: Dr. Herbert E. Rauch, Lockheed 92-20/205, Palo Alto Research Laboratory, 3251 Hanover Street, Palo Alto, CA 94304
Phone: 415-424-2704
ISSN: 0021-9142 **Circulation:** 1,500 **Frequency:** Q
Style Sheet: Yes
Revised Theses or Diss.: N/A **Manuscript Copies:** 4
Review Period: 1-2 months **Lagtime:** 3-6 months
Types of Articles: Research, reports, and reviews.
Major Contents: Astrodynamics, celestial mechanics, atmospheric flight mechanics, navigation of guidance, space-related sciences, altitude dynamics, orbit determination, trajectory optimization, space mission analysis, and numerical methods.
Miscellaneous: Issues are occasionally devoted to highlighting specific research area. For example, Volume 32, No.3 contained papers on Composite Materials in Space Applications.

344 **JOURNAL OF THE FRANKLIN INSTITUTE**

Pergamon Press, Ltd.
Headington Hill Hall
Oxford OX3 OBW
England

Editor: Lillian H. Falgie, Managing Editor, Journal of the Franklin Institute, 20th and Parkway, Philadelphia, PA 19103
Phone: 914-592-7700
ISSN: 0016-0032 **Circulation:** N/A **Frequency:** F
Style Sheet: "Suggestions to Authors."
Revised Theses or Diss.: Yes **Manuscript Copies:** 3
Review Period: 2 months (PR) **Lagtime:** 4-6 months
Types of Articles: Previously unpublished research, brief communications, and book reviews.
Major Contents: "Papers will be considered in all traditional branches of mathematics and the physical sciences, pure and applied, as well as interdisciplinary fields or composite sciences that combine the philosophies of two or more disciplines."
Miscellaneous: Articles of general interest will be given special attention and consideration.

345 **JOURNAL OF THE GEOLOGICAL SOCIETY**

Blackwell Scientific Publications, Ltd.
Osney Mead, Oxford
OX2 0EL, England

Editor: Dr. M.J. Le Bas, Publications Secretary, Geological Society, Burlington House, Piccadilly, London W1V 0JU, United Kingdom
Phone: 0533-555087
ISSN: 0016-7649 **Circulation:** 5,500 **Frequency:** BM
Style Sheet: Yes
Revised Theses or Diss.: Yes **Manuscript Copies:** 3
Review Period: 2-4 months (PR) **Lagtime:** 8 months
Types of Articles: Research, reviews, and theoretical.
Major Contents: Major geological topics of international interest in any of the earth sciences including plate tectonics, historical and evolutionary geology, geology of crust and mantle, oceans and ocean islands, structural geology, igneous metamorphic and sedimentary petrology, hydrogeology, processes of mineralization, and formation of oil and gas fields.

346 JOURNAL OF THE JAPANESE ASSOCIATION OF MINERALOGISTS PETROLOGISTS AND ECONOMIC GEOLOGISTS

Faculty of Science
Tohoku University
Sendai 980, Japan

Editor: Professor Satoshi Kanisawa
Phone: 81-0222-24-3852
ISSN: 0021-4825 **Circulation:** 1,300 **Frequency:** M
Style Sheet: Yes
Revised Theses or Diss.: Yes **Manuscript Copies:** 2
Review Period: 1-2 months (PR) **Lagtime:** 3-6 months
Types of Articles: Original articles, short notes, and book reviews.
Major Contents: Geochemistry, cosmochemistry, petrology, mineralogy, economic geology, organic geochemistry, sedimentology, volcanology, geothermal research, geochronology, cosmochronology, and meteoritics.

347 JOURNAL OF THE MECHANICS AND PHYSICS OF SOLIDS

Pergamon Press
Maxwell House, Fairview Park
Elmsford, NY 10523

Editor: H.G. Hopkins
Phone: 914-592-7700
ISSN: 0022-5096 **Circulation:** 1,500 **Frequency:** BM
Style Sheet: Available from the publisher.
Revised Theses or Diss.: N/A **Manuscript Copies:** 2

Review Period: 1 month (PR) **Lagtime:** 9 months
Types of Articles: Original papers, invited progress reports, and reviews.
Major Contents: Topics on the properties of constructional materials such as the connections between continuum and microstructural properties of materials.

348 JOURNAL OF THE WATER POLLUTION CONTROL FEDERATION

Water Pollution Control Federation
2626 Pennsylvania Avenue, N.W.
Washington, DC 20037

Editor: Peter J. Piecuch
Phone: 202-337-2500
ISSN: 0043-1303 **Circulation:** 32,000 **Frequency:** M
Style Sheet: Author's Guide in the journal.
Revised Theses or Diss.: No **Manuscript Copies:** 3
Review Period: 3 months (PR) **Lagtime:** 9 months
Types of Articles: Research, design, and operations.
Major Contents: Basic and practical knowledge concerning the technology of water pollution control, and the effects of pollution of natural waters.
Miscellaneous: An official publication of the Water Pollution Control Federation.

349 JOURNAL OF THERMAL INSULATION

Technomic Publishing Company, Inc.
265 Post Road, W.
Westport, CT 06880

Editor: Carlos J. Hilado
Phone: 717-291-5609
ISSN: 0148-8287 **Circulation:** 500 **Frequency:** Q
Style Sheet: Yes
Revised Theses or Diss.: N/A **Manuscript Copies:** 3
Review Period: 3 weeks **Lagtime:** N/A
Types of Articles: Research.
Major Contents: Thermal insulation techniques and products.

350 JOURNAL OF THERMAL STRESSES

Hemisphere Publishing Corporation
1010 Vermont Avenue, N.W.
Washington, DC 20005

Editor: Richard B. Hetnarski, Editor-in-Chief, Department of Mechanical Engineering, Rochester Institute of Technology, Rochester, NY 14623
Phone: 716-475-2157/2162
ISSN: 0149-5739 **Circulation:** 300 **Frequency:** Q
Style Sheet: Yes
Revised Theses or Diss.: N/A **Manuscript Copies:** 2
Review Period: 1-2 months (PR) **Lagtime:** 4-6 months
Types of Articles: Research, reviews, publications, and theoretical and experimental.
Major Contents: Topics include thermal stresses, new developments in thermoelasticity, thermoplasticity, approximate methods in thermal stresses, finite element method, thermodynamics, and heat transfer if related to thermal stresses.

351 JOURNAL OF TOXICOLOGY AND ENVIRONMENTAL HEALTH

Hemisphere Publishing Corporation
1010 Vermont Avenue, N.W.
Washington, DC 20005

Editor: Gunter Zweig and Renate D. Kimbrough, 4716 N. 20th Place, Arlington, VA 22207
Phone: 703-525-6282
ISSN: 0098-4108 **Circulation:** 1,246 **Frequency:** M
Style Sheet: Yes
Revised Theses or Diss.: Yes **Manuscript Copies:** 3
Review Period: N/A (PR) **Lagtime:** N/A
Types of Articles: Research, short research notes, reviews, symposia and conferences, and book reviews.
Major Contents: Places emphasis on the toxicological effect of natural and anthropogenic environmental pollutants and their action on intact organisms as well as in vitro systems. Fields of special interest are carcinogenesis, mutagenesis, teratology, neurotoxicity, environmental factors affecting health, and other toxicological phenomena, and epidemiological studies on select groups of workers or exposed populations are also of particular interest.

352 JOURNAL OF VOLCANOLOGY AND GEOTHERMAL RESEARCH

Elsevier Science Publishers B.V.
P.O. Box 211
1000 AE Amsterdam
The Netherlands

Editor: B.H. Baker and A.R. McBirney, Center for Volcanology, University of Oregon, Eugene, OR 97403
Phone: 212-916-1250
ISSN: 0377-0273 **Circulation:** N/A **Frequency:** M
Style Sheet: Yes
Revised Theses or Diss.: N/A **Manuscript Copies:** 3
Review Period: Varies **Lagtime:** Varies
Types of Articles: Research, reviews, short communications, letters, book reviews, and announcements.
Major Contents: Covers the following aspects of volcanology and geothermal research: (1) geochemical aspects geochemistry of volcanic rocks, geochemistry of volcanic gases and related products, isotope studies; (2) petrological aspects - magma genesis and magma evolution, petrology of volcanic rocks; (3) economic aspects - ore deposits related to volcanic rocks, geothermal energy; (4) geophysical aspects - physical properties of volcanic rocks and magmas, heat flow studies, physical volcanology, volcanic seismology, volcanic surveillance, geophysical exploration for geothermal resources; (5) tectonic aspects - relation of volcanism to global and regional tectonic processes; (6) environmental aspects - volcanic gases and their effect on the atmosphere.

353 KERNENERGIE ZEITSCHRIFT FUR KERNFORSCHUNG UND KERNTECHNIK

Akademie Verlag Berlin
DDR 1086 Berlin
Leipziger Strasse 3/4
Germany

Editor: Professor Dr. Gunter Schumann
Phone: Zittau 610
ISSN: 0023-0642 **Circulation:** N/A **Frequency:** M
Style Sheet: No
Revised Theses or Diss.: Yes **Manuscript Copies:** 2
Review Period: 1 month **Lagtime:** 8 months
Types of Articles: General and specialized articles dealing with all areas of nuclear power, engineering and technology, also reviews, letters to the editor, proceedings of meetings, and book reviews.
Major Contents: Nuclear power plants; nuclear reactor technology; nuclear fuels; physics research (cont.); radiation and shielding physics, nuclear physics; fusion energy; fusion power plant technology; energy planning and policy; and nuclear energy.

354 **L'ENERGIA ELETTRICA**

Associazione Elettrotecnica Ed Elettronica Italiana
Viale Monza 259
20126 Milan, Italy

Editor: Giorgio Catenacci
Phone: 25-50-641
ISSN: 0013-7308 **Circulation:** 4,600 **Frequency:** M
Style Sheet: Yes
Revised Theses or Diss.: Yes **Manuscript Copies:** 3
Review Period: 2 months (PR) **Lagtime:** 2 months
Types of Articles: Research, projects, and reviews.
Major Contents: All aspects of electric energy.
Miscellaneous: Primarily Italian. However, selected articles are in English.

355 **LAND ECONOMICS**

University of Wisconsin, Taylor Hall
427 Lorch Street, Room 109
Madison, WI 53706

Editor: Daniel Bromley
Phone: 608 262 2400
ISSN: 0023-7639 **Circulation:** 3,200 **Frequency:** Q
Style Sheet: Yes
Revised Theses or Diss.: No **Manuscript Copies:** 3
Review Period: 2 months (PR) **Lagtime:** 6-8 months
Types of Articles: Research.
Major Contents: Conceptual and empirical research concerning land, natural resources, conservation, energy, urban economics, housing, and transportation.

356 **LASER FOCUS: THE MAGAZINE OF ELECTRO-OPTICS TECHNOLOGY**

PennWell Publishing Company
119 Russell Street
Littleton, MA 01460

Editor: Richard R. Mack, Managing Editor
Phone: 617-486-9501
ISSN: 8755-1853 **Circulation:** 41,000 **Frequency:** M
Style Sheet: Yes
Revised Theses or Diss.: Yes **Manuscript Copies:** 2

Review Period: 2 months (PR) **Lagtime:** 3-6 months
Types of Articles: Research, projects, reviews, and case studies.
Major Contents: Papers on a mixture of topics, including basic research, R&D and systems design work including focus on newsbreaking developments in lasers, laser applications, infrared technology, optics, fiberoptics, and other areas of electro-optics.
Miscellaneous: This is an industry news magazine, not a research journal. However, contributed reviews and technical articles are published on a regular basis.

357 LATIN AMERICAN ENERGY UPDATE
Greater Columbia Magazine Company
1120 Broad River Road
Columbia, SC 29210

Editor: Joao Pimentel, c/o OLADE, P.O. Box 6413 C.C.I., Quito, Ecuador
Phone: 538-785
ISSN: N/A **Circulation:** 2,000 **Frequency:** IR
Style Sheet: No
Revised Theses or Diss.: No **Manuscript Copies:** 1
Review Period: 3-4 months **Lagtime:** 4-8 months
Types of Articles: News articles.
Major Contents: News on the Latin American energy scene including new and renewable sources of energy, non-renewable sources of energy, energy planning and conservation, energy economics, and energy technology.

358 LITHOS: AN INTERNATIONAL JOURNAL OF MINERALOGY, PETROLOGY AND GEOCHEMISTRY
Elsevier Science Publishers B.V.
P.O. Box 330
1000 AH Amsterdam, The Netherlands

Editor: W.L. Griffin and E.R. Neumann, Mineralogisk-Geologisk Museum, Sars Gate 1, Oslo 5, Norway
Phone: 020-5803-911
ISSN: 0024-4937 **Circulation:** 800 **Frequency:** Q
Style Sheet: Yes
Revised Theses or Diss.: Yes **Manuscript Copies:** 3
Review Period: Varies **Lagtime:** Varies
Types of Articles: Scholastic papers, reviews, discussions, and book reviews.
Major Contents: Topics of international interest in the fields of mineralogy, petrology, and geochemistry.

359 **LOG ANALYST, THE**

Society of Professional Well Log Analysts
806 Main Street, Suite 1017
Houston, TX 77002

Editor: Harry Brown, Editor
Phone: 713-871-9440
ISSN: 0024-581X **Circulation:** 4,000 **Frequency:** BM
Style Sheet: New York Times.
Revised Theses or Diss.: Yes **Manuscript Copies:** 4
Review Period: 2 months **Lagtime:** 2 months
Types of Articles: Research, reviews, case histories, new applications, and new techniques.
Major Contents: Formation evaluation and oil and gas drilling applications and techniques.

360 **LONG RANGE PLANNING**

Pergamon Press, Ltd.
Headington Hill Hall
Oxford, 0X3 0BW
England

Editor: Professor Bernard Taylor, Henley-The Management College, Greenlands, Henley-on-Thames, Oxon RG9 3AU, United Kingdom
Phone: Hambleden 0491-571454
ISSN: 0024-6301 **Circulation:** N/A **Frequency:** BM
Style Sheet: Yes
Revised Theses or Diss.: N/A **Manuscript Copies:** 3
Review Period: N/A (PR) **Lagtime:** N/A
Types of Articles: Research, book reviews, and recent developments.
Major Contents: Concepts and techniques involved in the development and implementation of strategy and plans.
Miscellaneous: The Journal of the Society for Long Range Planning. American editor is Ian Wilson, SRI International, 333 Ravenswood Avenue, Menlo Park, CA 94025, 914-592-7700.

361 **LUBRICATION ENGINEERING**

American Society of Lubrication Engineers
838 Busse Highway
Park Ridge, IL 60068

Editor: J. McCoy
Phone: 312-825-5536

ISSN: 0024-7154 **Circulation:** 6,400 **Frequency:** M
Style Sheet: ASLE Author's Guide.
Revised Theses or Diss.: Yes **Manuscript Copies:** 8
Review Period: 1-2 months (PR) **Lagtime:** 12 months
Types of Articles: Design, research, applications, education, and maintenance.
Major Contents: Lubrication in all of its branches including friction and wear, bearings, seals, hydrodynamic lubrication, engine lubrication, metalworking, food, drugs and cosmetics, hydraulics, petroleum and chemicals, steel, mining, nonferrous metals, lubrication equipment and practices, power generation, lubrication fundamentals, and aerospace.

362 **MARINE MINING**
Crane, Russak and Company, Inc.
3 E. 44th Street
New York, NY 10017

Editor: J. Robert Moore
Phone: 212-867-1490
ISSN: 0149-0397 **Circulation:** N/A **Frequency:** Q
Style Sheet: Yes
Revised Theses or Diss.: N/A **Manuscript Copies:** 3
Review Period: 2-4 months **Lagtime:** 6-10 months
Types of Articles: Research and reviews.
Major Contents: Undersea mining, mining, ocean sciences and engineering.

363 **MARINE TECHNOLOGY SOCIETY JOURNAL**
The Marine Technology Society
1730 M Street, N.W., Suite 412
Washington, DC 20036

Editor: Rita Colwell
Phone: 202-659-3251
ISSN: 0025-3324 **Circulation:** 5,000 **Frequency:** Q
Style Sheet: Yes
Revised Theses or Diss.: N/A **Manuscript Copies:** 3
Review Period: Varies **Lagtime:** 6 months
Types of Articles: Papers and technical notes.
Major Contents: All aspects of marine technology.

364 **MATERIALS AND COMPONENTS IN FOSSIL ENERGY APPLICATIONS**
Battelle Columbus Laboratories, Corrosion Section
505 King Avenue,
Columbus, OH 43201

Editor: Mr. E.E. Hoffman
Phone: 615-576-0735
ISSN: 0145-9244 **Circulation:** 1,300 **Frequency:** BM
Style Sheet: No
Revised Theses or Diss.: No **Manuscript Copies:** 1
Review Period: 1 month **Lagtime:** 1-2 months
Types of Articles: Research in progress, project descriptions, and reviews.
Major Contents: Articles concerned with the areas of materials application and performance, and component design and performance in processes involving the combustion, gasification, or liquefaction of coal. Properties, corrosion behavior, wear performance of alloys, refractories, ceramics, and other non-metals; failure analyses; reports of good performance in difficult applications, component design (pumps, valves, cyclones, turbines, transfer lines, heat exchangers, pressure vessels, combustors, etc.), reports of performance, welding techniques, techniques for measurement of corrosion deposition, and erosion deposition.
Miscellaneous: Send manuscript to Dr. I.G. Wright, (Journal's Address).

365 **MATERIALS AND SOCIETY**
Pergamon Press
Maxwell House, Fairview Park
Elmsford, NY 10523

Editor: Walter R. Hubbard, Jr., Acta Metallurgica, Box 8, Schenectady, NY
Phone: 914-592-7700
ISSN: 0146-6399 **Circulation:** 1,000 **Frequency:** Q
Style Sheet: Yes
Revised Theses or Diss.: N/A **Manuscript Copies:** N/A
Review Period: 1 month (PR) **Lagtime:** 6 months
Types of Articles: Research and short communications.
Major Contents: Materials availability; energy and economics issues concerning the interaction of materials and society.

366 **MATERIALS SCIENCE AND ENGINEERING**
Elsevier Sequoia S.A.
P.O. Box 851
CH-1001 Lausanne 1
Switzerland

Editor: Robert Maddin

ISSN: 0025-5416 **Circulation:** N/A **Frequency:** M
Style Sheet: Yes
Revised Theses or Diss.: N/A **Manuscript Copies:** 3
Review Period: 1 month **Lagtime:** 6-8 months
Types of Articles: Research, reviews, communications, and letters to the editor.
Major Contents: Properties of materials, especially pertaining to their structure and use in engineering.

367 **MECHANICAL ENGINEERING**

American Society of Mechanical Engineers
345 E. 47th Street
New York, NY 10017

Editor: Charles W. Beardsley
Phone: 212-705-7782
ISSN: 0025-6501 **Circulation:** 115,000 **Frequency:** M
Style Sheet: An ASME Paper, ASME Manual MS-4.
Revised Theses or Diss.: Yes **Manuscript Copies:** 2
Review Period: 2-3 months (PR) **Lagtime:** 6-12 months
Types of Articles: Research, projects, and reviews.
Major Contents: Covers a wide range of technologies including materials development, aerospace industries, turbomachinery, computer applications, and photoelasticity. Each issue normally contains one energy related article.
Miscellaneous: An official journal of the American Society of Mechanical Engineers.

368 **MEMBRANE SCIENCE AND DESALINATION**

Elsevier Science Publishers B.V.
P.O. Box 330
1000 AH Amsterdam
The Netherlands

Editor: M. Balaban
Phone: 020-5803-911
ISSN: 0011-9164 **Circulation:** N/A **Frequency:** 21/year
Style Sheet: Yes
Revised Theses or Diss.: N/A **Manuscript Copies:** 2
Review Period: 1-2 months (PR) **Lagtime:** 5 months
Types of Articles: Research, reviews, and technical papers.
Major Contents: Theoretical and applied research, technological and industrial development, and experience of operators and users in the water desalting and purification field. Also includes all the desalting fields

such as: distillation; membranes; reverse osmosis; electrodialysis; ion exchange; freezing; and water purification.
Miscellaneous: Formed by the merger of: JOURNAL OF MEMBRANE SCIENCE, ISSN: 0376-7388 and DESALINATION, ISSN: 0011-9164.

369 **MERCIAN GEOLOGIST**
Department of Geology
The University of Nottingham
Nottingham NG7 2RD
England

Editor: R.J. Firman
Phone: 0602-267-442
ISSN: 0025-990X **Circulation:** 2,740 **Frequency:** SA
Style Sheet: Yes
Revised Theses or Diss.: Yes **Manuscript Copies:** 2
Review Period: 2-3 months **Lagtime:** 6 months
Types of Articles: Research, reviews, and excursion reports.
Major Contents: Geology of the English Midlands, and current interest articles concerning geology in general.
Miscellaneous: Journal of the East Midlands Geological Society.

370 **METALLURGICAL TRANSACTIONS**
Metallurgical Society of America
420 Commonwealth Drive, Box 430
Warrendale, PA 15086

Editor: Edwin Ruh
Phone: 216-388-5151
ISSN: 0360-2133 **Circulation:** 3,546 **Frequency:** M
Style Sheet: Yes
Revised Theses or Diss.: N/A **Manuscript Copies:** 3
Review Period: 3 months **Lagtime:** 6 months
Types of Articles: Research.
Major Contents: Topics in materials science, metallurgy, and the processing of metals including mineral benefication and preparation, hydro-, pyro-, and electro-metallurgy, process models, and related computer applications.
Miscellaneous: An official publication of the Metallurgical Society of American Institute of Mining, Metallurgical, and Petroleum Engineers, Inc.

371 MICHIGAN ENERGY JOURNAL

Michigan Solar Energy Association
415 Detroit Street
Ann Arbor, MI 48104

Editor: Howard Meyerson
Phone: 313-668-0555
ISSN: 0748-2388 **Circulation:** 1,500 **Frequency:** BM
Style Sheet: Yes
Revised Theses or Diss.: Yes **Manuscript Copies:** 2
Review Period: 1 month **Lagtime:** 1 month
Types of Articles: Technical reports, research, case studies, service articles, profiles (business and personal), political analysis editorials, general information related to solar and renewable energy efficiency, business tips, financing, comparatives on various systems, and software.
Major Contents: Any solar area from technological, political, economic, practical, system comparisons, and useful information for homeowners and builders.
Miscellaneous: Revised theses and dissertations are accepted only following verbal contact. Unsolicited manuscripts are accepted; however, queries or phone calls are preferred.

372 MICROBIOLOGICAL REVIEWS

American Society for Microbiology
1913 I Street, N.W.
Washington, DC 20006

Editor: Howard V. Rickenberg, Division of Molecular and Cellular Biology, National Jewish Hospital and Research Center, Denver, CO 80206
Phone: 202-833-9680
ISSN: 0146-0749 **Circulation:** 16,684 **Frequency:** Q
Style Sheet: Yes
Revised Theses or Diss.: N/A **Manuscript Copies:** N/A
Review Period: 1-3 months (PR) **Lagtime:** N/A
Types of Articles: Reviews and monographs.
Major Contents: All aspects of microbiology.
Miscellaneous: Formerly: BACTERIOLOGICAL REVIEWS.

373 MICROCOMPUTER APPLICATIONS

Acta Press
P.O. Box 25, Station G
Calgary, Alberta
T3A 2G1, Canada

Editor: G.K.F. Lee
Phone: 403-270-3616
ISSN: 0820-0750 **Circulation:** 500 **Frequency:** 3/year
Style Sheet: No
Revised Theses or Diss.: Yes **Manuscript Copies:** 3
Review Period: 2 months (PR) **Lagtime:** N/A
Types of Articles: Book reviews, conferences, call for papers, new publications, advertisements, and research.
Major Contents: All aspects of microcomputer applications including engineering, science, business, management, robotics, manufacturing, education, and personal computers.
Miscellaneous: A publication of the International Society for Mini and Microcomputers-ISMM.

374 **MICRON**

Pergamon Press
Maxwell House, Fairview Park
Elmsford, NY 10523

Editor: Professor R.W. Horne
Phone: 914-592-7700
ISSN: 0047-7206 **Circulation:** 1,100 **Frequency:** Q
Style Sheet: Yes
Revised Theses or Diss.: N/A **Manuscript Copies:** 2
Review Period: 1 month **Lagtime:** 6-7 months
Types of Articles: Research, communications, reviews, conferences, and reports.
Major Contents: Electron microscopy and electron probe micro-analysis.

375 **MINERAL AND ENERGY RESOURCES**

Colorado School of Mines Press Publications Center
Chauvenet Hall
Golden, CO 80401

Editor: Jon W. Raese
Phone: 303-273-3600
ISSN: 0192-6179 **Circulation:** 1,382 **Frequency:** BM
Style Sheet: Yes
Revised Theses or Diss.: Yes **Manuscript Copies:** 2
Review Period: 1 month (PR) **Lagtime:** 3 months
Types of Articles: Topical, technical or critical articles, and reviews.
Major Contents: Technology forecasting in minerals and energy and articles on mineral economics. All the earth sciences including geology,

geophysics, metallurgy, mining, petroleum, chemical and petroleum refining, chemistry and geochemistry, environmental science, mineral economics, mineral engineering, energy management, engineering education, and resource studies in minerals and energy.

Miscellaneous: Letters are requested prior to the submission of material. Two reviews from competent critics, preferably from outside the author's institution , must accompany each submission. Critical reviews must be thoroughly documented assessments.

376 **MINES MAGAZINE**

Colorado School of Mines Press Publications Center
Chauvenet Hall
Golden, CO 80401

Editor: Patricia Petty
Phone: 303-273-3293
ISSN: 0096-4859 **Circulation:** 7,000 **Frequency:** M
Style Sheet: Yes
Revised Theses or Diss.: Yes **Manuscript Copies:** 1
Review Period: 2 months **Lagtime:** 1 month
Types of Articles: Research, projects, and reviews.
Major Contents: Mineral policy, political and world issues, business, and technological developments affecting the nation's basic industries: mining, geophysics, hazardous materials, petroleum, synfuels, geology, and computers.

377 **MINING SCIENCE AND TECHNOLOGY**

Elsevier Science Publishers B.V.
P.O. Box 330
1000 AH Amsterdam
The Netherlands

Editor: C.O. Brawner, Department of Mining and Mineral Process, University of British Columbia, Forward Building, Vancouver, BC V6T 1W5, Canada
Phone: 604-228-3986
ISSN: 0167-9031 **Circulation:** N/A **Frequency:** Q
Style Sheet: Yes
Revised Theses or Diss.: Yes **Manuscript Copies:** 3
Review Period: 3 months (PR) **Lagtime:** 3 months
Types of Articles: Theoretical articles, reviews of research, accounts of mining practice, and selected case histories.
Major Contents: All aspects of the extraction from the earth of solid-mineral materials including exploration and mine economics, mine plan-

ning and design, surface and underground excavation methods and equipment, ground control, mine drainage, mine surveying, mine transport systems, ventilation, environmental control, waste disposal, mine safety, legal aspects of mining, and future mining.

Miscellaneous: Official language is English but occasional articles in French or German will be accepted.

378 **MISSISSIPPI GEOLOGY**

Department of Natural Resources, Bureau of Geology
2525 Northwest Street, Box 5348
Jackson, MS 39216

Editor: Michael B.E. Bograd and David Dockery
Phone: 601-354-6228
ISSN: 0275-8555 **Circulation:** 1,776 **Frequency:** Q
Style Sheet: Yes
Revised Theses or Diss.: Yes **Manuscript Copies:** 1
Review Period: 2 months **Lagtime:** 6 months
Types of Articles: Short research, projects, letters, and reviews.
Major Contents: Topics pertaining to the geology of Mississippi including general geology, petroleum geology, stratigraphy, petrology, mineralogy, paleontology, and mineral resources (fuel and nonfuel).

379 **MODERN GEOLOGY**

Gordon and Breach Science Publishers, Ltd.
1 Bedford Street
London WC2E 9TP
England

Editor: B. Halstead, Department of Geology, University of Reading, Reading RG6 2AB, United Kingdom
Phone: 202-206-8900
ISSN: 0026-7775 **Circulation:** N/A **Frequency:** Q
Style Sheet: Yes
Revised Theses or Diss.: N/A **Manuscript Copies:** 2
Review Period: 2-3 months **Lagtime:** 2 months
Types of Articles: State-of-the-art articles and reviews, original papers, and short communications.
Major Contents: Broad fields of physical, chemical, mathematical, statistical, and extraterrestrial geology.

380 **MODERN POWER AND ENGINEERING**

Maclean-Hunter, Ltd.
777 Bay Street
Toronto, Ontario
M5N 1A7, Canada

Editor: Bruce Glassford
Phone: 416-596-5000
ISSN: N/A **Circulation:** 14,000 **Frequency:** M
Style Sheet: N/A
Revised Theses or Diss.: N/A **Manuscript Copies:** N/A
Review Period: N/A **Lagtime:** N/A
Types of Articles: Technological advancements, industrial applications, and equipment reviews.
Major Contents: Services the field of plant power services from generation to utilization. Topics relate to electricity, steam, water, compressed air, refrigeration and air conditioning.

381 MOLECULAR AND CELLULAR BIOLOGY
ASM Publications Department
1913 I Street, N.W.
Washington, DC 20006

Editor: Aaron J. Shatkin, Roche Institute of Molecular Biology, Nutley, NJ
Phone: 202-833-9680
ISSN: 0270-7306 **Circulation:** 4,000 **Frequency:** M
Style Sheet: January issue of journal.
Revised Theses or Diss.: Yes **Manuscript Copies:** N/A
Review Period: N/A (PR) **Lagtime:** N/A
Types of Articles: Research.
Major Contents: Fundamental knowledge concerning the molecular biology of eucaryotic cells, of both microbial and higher organisms.
Miscellaneous: An official publication of the American Society for Microbiology.

382 MONTANA OIL JOURNAL
Montana Oil Journal, Inc.
P.O. Box 20255
Billings, MT 59104

Editor: Roy Boles
Phone: 406-656-0886
ISSN: 0047-794X **Circulation:** 5,800 **Frequency:** W
Style Sheet: No
Revised Theses or Diss.: No **Manuscript Copies:** 1
Review Period: 1 week **Lagtime:** 1 week
Types of Articles: Research, news articles, and informational highlights.

Major Contents: Oil and gas activity. Limited to Montana and Williston basin exploration and production.

383 **NATURAL GAS**
Benn Publications
Sovereign Way, Tonbridge
Kent TN9 1RW, England

Editor: Victoria Thomas
Phone: 0732-362468
ISSN: 0140-3222 **Circulation:** 28,773 **Frequency:** BM
Style Sheet: Yes
Revised Theses or Diss.: No **Manuscript Copies:** 1
Review Period: 2 months **Lagtime:** 1 month
Types of Articles: Research, projects, and general interest articles relating to natural gas.
Major Contents: Gas utilization in industry and commerce, gas and energy efficiency technology, gas availability and the marketing of future supplies, and recent developments in plant, equipment and processes.
Miscellaneous: Articles are not too technical and should be of interest to the business community.

384 **NATURAL RESOURCES FORUM**
Graham and Trotman, Ltd.
Sterling House, 66 Wilton Road
London SWIV 1DE
England

Editor: Raymond Knowles, Managing Editor, Natural Resources Forum, United Nations, 1 UN Plaza, United Nations, New York, NY 10017
Phone: 212-754-1234
ISSN: 0165-0203 **Circulation:** 1,000 **Frequency:** Q
Style Sheet: *See Miscellaneous
Revised Theses or Diss.: Yes **Manuscript Copies:** 3
Review Period: 1-2 months (PR) **Lagtime:** 3-6 months
Types of Articles: Research, projects, reviews, comprehensive studies, and analyses.
Major Contents: Major economic, technological, legal, financial and policy issues associated with energy, mineral and water resources in developing countries. Factors affecting exploration, development, management and utilization of these resources in developing countries.
Miscellaneous: Journal audience includes Government policy-makers, planners, researchers, bankers, mineral and energy company officials. *Style guide is in Graham and Trotman Journals, Instructions to Con-

tributors.

385 NATURAL RESOURCES JOURNAL

The University of New Mexico
School of Law
1117 Stanford, N.W.
Albuquerque, NM 87131

Editor: Albert E. Utton
Phone: 505-277-2146
ISSN: 0028-0739 **Circulation:** 2,000 **Frequency:** Q
Style Sheet: Yes
Revised Theses or Diss.: N/A **Manuscript Copies:** 3
Review Period: N/A **Lagtime:** N/A
Types of Articles: Research, book reviews, and notes.
Major Contents: Topics on natural and environmental resources which emphasize public policy.

386 NATURE: THE INTERNATIONAL JOURNAL OF SCIENCE

Macmillan Journals, Inc.
1134 National Press Building
Washington, DC 20045

Editor: John Maddox, 4 Little Essex Street, London L5C2R 3LF, England
Phone: 01-836-6633
ISSN: 0028-0836 **Circulation:** 27,092 **Frequency:** W
Style Sheet: Yes
Revised Theses or Diss.: No **Manuscript Copies:** 3
Review Period: 1 month (PR) **Lagtime:** 3 months
Types of Articles: Reviews, research, letters, and brief comments.
Major Contents: Current science. Current R&D in all sciences, and international general science.

387 NEUES JAHRBUCH FUR GEOLOGIE UND PALAONTOLOGIE MONATSHEFTE

E. Schweizerbart'sche
Johannesstrasse 3A, D-7000
Stuttgart 1
West Germany

Editor: Professor Dr. John G. Dennis, California State University, 1250 Bellflower Boulevard, Long Beach, CA 90840
Phone: 213-498-4404

ISSN: 0028-3630 **Circulation:** N/A **Frequency:** M
Style Sheet: Yes
Revised Theses or Diss.: Yes **Manuscript Copies:** 2
Review Period: 1 month **Lagtime:** 6 months
Types of Articles: Research and projects.
Major Contents: General and applied geology including geology of mineral deposits, historical and regional geology, paleozoology, paleobotany, general paleontology, and biostratigraphy.
Miscellaneous: All articles are to be preceded by a brief "Zusammenfussuag" in German and a corresponding "Abstract" in English. If the article is not in English, an English translation is desirable.

388 NOROIL: THE INTERNATIONAL OIL AND GAS MAGAZINE

P.O. Box 480
Hillevagsun. 17
4001 Stavanger
Norway

Editor: Ole Jacob Kvinnsland
Phone: 04-58-90-00
ISSN: 0332-544X **Circulation:** 14,550 **Frequency:** M
Style Sheet: Yes
Revised Theses or Diss.: Yes **Manuscript Copies:** 2
Review Period: 2 weeks **Lagtime:** 1 month
Types of Articles: Technical papers and reviews.
Major Contents: Offshore industry technology including offshore inspection, maintenance and repair, specialized vessels, advances in offshore technology, drilling ing and well completion, mobile rigs report, offshore pipelines, subsea technology, electronic/computers offshore, offshore field development, offshore finance and insurance, and LPG-LNG trade and technology.

389 NORSK GEOLOGISK TIDSSKRIFT: A NORWEGIAN JOURNAL OF GEOLOGY

Ndorges Geologiske Undersokelse
Postboks 3006 N-7001 Trondheim
Norway

Editor: Gunnar Juve
ISSN: 0029-196X **Circulation:** 1,350 **Frequency:** Q
Style Sheet: Yes
Revised Theses or Diss.: Yes **Manuscript Copies:** 3
Review Period: 2-4 months **Lagtime:** 2-6 months
Types of Articles: Research and projects.

Major Contents: All geological activity in Norway and adjacent areas including petrography, petrology, mineralogy, paleontology, tectonics, stratigraphy, and quarternary geology.
Miscellaneous: Articles should be submitted preferably in English; however, Norwegian, Danish, Swedish, French or German may be used.

390 **NUCLEAR ENERGY**
Thomas Telford, Ltd.
1-7 Great George Street
Westminister
London SWIP 3AA, England

Editor: Dr. J.L. Head
Phone: London 01-630-0726
ISSN: 0140-4067 **Circulation:** 3,000 **Frequency:** BM
Style Sheet: Yes
Revised Theses or Diss.: Yes **Manuscript Copies:** 3
Review Period: 1-2 months (PR) **Lagtime:** 6 months
Types of Articles: Research, reviews, projects, news articles and lectures.
Major Contents: All areas of concern to nuclear scientists and engineers.
Miscellaneous: An official publication of the British Nuclear Energy Society. Formerly: BRITISH NUCLEAR ENERGY SOCIETY JOURNAL, ISSN: 0007-1587.

391 **NUCLEAR ENGINEER, THE**
The Institution of Nuclear Engineers
Allan House
1 Penerley Road
London SE6 2LQ, England

Editor: Mrs. S.M. Blackburn
Phone: 01-698-1500
ISSN: 0262-5091 **Circulation:** 2,000 **Frequency:** BM
Style Sheet: Yes
Revised Theses or Diss.: No **Manuscript Copies:** 3
Review Period: (PR) **Lagtime:** 9 months
Types of Articles: Research.
Major Contents: All aspects of nuclear power generation, research, marine propulsion, health and environmental topics as well as general industrial uses and applications. In the United Kingdom it provides a vehicle for manufacturers to reach all concerned with design, manufacture and installation of whole process plants as well as individual items of related nucleonics equipment.

392 **NUCLEAR ENGINEERING AND DESIGN**

Elsevier Science Publishers B.V.
P.O. Box 211
1000 AE Amsterdam
The Netherlands

Editor: Stanley H. Fistedis, Principal Editor, Engineers Mechanics Programs, Argonne National Laboratory, Argonne, IL 60439
Phone: 312-972-4700
ISSN: 0029-5493 **Circulation:** 1,800 **Frequency:** 384
Style Sheet: Yes
Revised Theses or Diss.: Yes **Manuscript Copies:** 3
Review Period: 2-6 months (PR) **Lagtime:** 6-7 months
Types of Articles: Research, projects, and letters to the editor.
Major Contents: Thermal, mechanical materials, structural problems of nuclear fission energy, nuclear reactor systems and safety, fluids, thermohydraulics, heat transfer, and structural materials pertinent to fission reactor engineering and design.

393 **NUCLEAR ENGINEERING INTERNATIONAL**

Business Press International
Quadrant House, The Quadrant
Sutton, Surrey SM2 5AS
England

Editor: James Varley
Phone: 01-661-3318
ISSN: 0029-5507 **Circulation:** 3,481 **Frequency:** M
Style Sheet: No
Revised Theses or Diss.: N/A **Manuscript Copies:** N/A
Review Period: 1 month **Lagtime:** N/A
Types of Articles: News, comments, and research articles.
Major Contents: All aspects of nuclear engineering including power plant siting and construction; decommissioning projects; reactor safety, reliability, performance, and monitoring; fuel elements and assemblies; control and surveillance of reactors of any type; environmental aspects of nuclear power plants; and economic analyses related to various aspects of nuclear projects.
Miscellaneous: Articles, while technical, are informative and useful to practicing nuclear engineers. Theoretical articles are not generally accepted.

394 **NUCLEAR FUSION**

International Atomic Energy Agency
Wagramerstrasse 5
P.O. Box 100
A-1400 Vienna, Austria

Editor: Mrs. Dorianna Twersky
Phone: Austria 222-2360-2871
ISSN: 0029-5515 **Circulation:** 2,000 **Frequency:** M
Style Sheet: Yes
Revised Theses or Diss.: Yes **Manuscript Copies:** 3
Review Period: 3-4 months (PR) **Lagtime:** 3 months
Types of Articles: Research, conceptual studies, solicited review papers, major conference summaries, comments, and replies.
Major Contents: Basic and applied fusion plasma physics, plasma engineering, divertors, plasma heating systems, experimental devices, design and operation of fusion experiments, diagnostics, conceptual design studies of reactors, and inertial confinement fusion.
Miscellaneous: Publishes a regular triennial supplement to Nuclear Fusion, "World Survey of Major Activities in CTR" and other series.

395 **NUCLEAR MATERIALS MANAGEMENT**

Institute of Nuclear Materials Management, Inc.
8600 W. Bryn Mawr Avenue
Chicago, IL 60631

Editor: John E. Messervey
Phone: 312-693-0990
ISSN: 0362-0034 **Circulation:** 760 **Frequency:** Q
Style Sheet: Yes
Revised Theses or Diss.: Yes **Manuscript Copies:** 1
Review Period: 1 month **Lagtime:** 3-6 months
Types of Articles: Research, projects, reviews, news items, comments, and reports on committee and chapter activities.
Major Contents: Advances in nuclear materials management and safeguards; all aspects of the fuel cycle including fuel fabrication, reactor, transportation, and waste management safeguards.

396 **NUCLEAR NEWS**

American Nuclear Society
555 N. Kensington Avenue
La Grange Park, IL 60525

Editor: Jon Payne

Phone: 312-352-6611
ISSN: 0029-5574 **Circulation:** 18,000 **Frequency:** M
Style Sheet: Yes
Revised Theses or Diss.: No **Manuscript Copies:** 1
Review Period: 1-2 months **Lagtime:** 1-3 months
Types of Articles: Overview (survey), projects, and non-commercial applications.
Major Contents: Applications of nuclear energy, principally in the generation of electricity including current problems and solutions in nuclear plant licensing, design, maintenance techniques, personnel training, and health physics.

397 **NUCLEAR SCIENCE AND ENGINEERING**
American Nuclear Society
555 N. Kensington Avenue
La Grange Park, IL 60525

Editor: Dr. John M. Kallfelz, Nuclear Engineering & Health Physics Program, Cherry Emerson Building, Georgia Institute of Technology, Atlanta, Georgia 30332
Phone: 404-894-3746
ISSN: 0029-5639 **Circulation:** 1,500 **Frequency:** M
Style Sheet: API Style Manual.
Revised Theses or Diss.: Yes **Manuscript Copies:** N/A
Review Period: 2-4 months (PR) **Lagtime:** 2-4 months
Types of Articles: Technical papers and notes, critical reviews, book reviews, computer code abstracts, and letters to the editor.
Major Contents: Topics related to the peaceful utilization of nuclear energy, radiation, and alternate energy sources.
Miscellaneous: The research journal of the American Nuclear Society. Manuscripts originating in Europe should be sent to Dr. Gunther Kessler, Associate Editor, Nuclear Science and Engineering, Kernforschugszentrum Karlsruhe, Postfachj 3640, D-7500 Karlsruhe 1, FRG. Phone: 07247-822440.

398 **NUCLEAR SCIENCE JOURNAL**
National Tsing Hua University
c/o Institute of Nuclear Science
855 Kuang Fu Road
Hsinchu, Taiwan 300, Rep. of China

Editor: Professor Pao-Shan Weng, Editor-in-Chief
Phone: 035-718649

ISSN: 0029-5647 **Circulation:** 2,000 **Frequency:** Q
Style Sheet: Yes
Revised Theses or Diss.: Yes **Manuscript Copies:** 2
Review Period: 1 month **Lagtime:** 1 month
Types of Articles: Research and reviews.
Major Contents: Nuclear science, nuclear engineering, nuclear technology, and related fields.
Miscellaneous: Articles are in English and Chinese. The Journal of the Atomic Energy Council, Republic of China.

399 NUCLEAR TECHNOLOGY

American Nuclear Society
555 N. Kensington Avenue
La Grange Park, IL 60525

Editor: Dr. Roy G. Post, Nuclear Technology, Department of Nuclear and Energy Engineering, The University of Arizona, Tucson, AZ 85721
Phone: 602-621-6158
ISSN: 0029-5450 **Circulation:** 2,000 **Frequency:** M
Style Sheet: AIP Style Manual.
Revised Theses or Diss.: Yes **Manuscript Copies:** 3
Review Period: 1 month (PR) **Lagtime:** 6 months
Types of Articles: Technical articles, technical notes, critical reviews, letters to the editor, and book reviews.
Major Contents: Fission reactors, nuclear safety, fuel cycles, chemical processing, nuclear fuels, radioactive waste management, economics, materials, radioisotopes and isotope separation, radiation biology and environment, heat transfer and fluid flow, criticality safety, analyses, techniques, and education.

400 NUCLEAR TRACKS AND RADIATION MEASUREMENTS

Pergamon Press, Ltd.
Headington Hill Hall
Oxford OX3 OBW
England

Editor: S.A. Durrani, Department of Physics, University of Birmingham, P.O. Box 363, Birmingham B15 2TT, England
Phone: 021-472-1301
ISSN: 0191-278X **Circulation:** 1,000 **Frequency:** Q
Style Sheet: Yes
Revised Theses or Diss.: No **Manuscript Copies:** 3
Review Period: 1 month (PR) **Lagtime:** 6 months
Types of Articles: Original papers, short communications, reviews, and

survey articles.

Major Contents: Radiation measurements, nuclear tracks, thermoluminescence, and related phenomena; cosmic-ray studies; space-mission based observations; energy transfer phenomena, especially in relation to biological systems; radiation protection and monitoring.

Miscellaneous: The publication aims to present an outlet for quality research of most recent development in the defined field. Editors in Eastern and Western Europe are identified in issues of the journal.

401 NUMERICAL FUNCTIONAL ANALYSIS AND OPTIMIZATION

Marcel Dekker, Inc.
270 Madison Avenue
New York, NY 10016

Editor: M.Z. Nashed, Editor-in-Chief, Department of Mathematical Sciences, University of Delaware, Newark, DE 19711
Phone: 212-696-9000
ISSN: 0163-0563 **Circulation:** N/A **Frequency:** SA
Style Sheet: Yes
Revised Theses or Diss.: N/A **Manuscript Copies:** 3
Review Period: 2-6 months (PR) **Lagtime:** 7 months

Types of Articles: Reviews, annotated bibliographies, surveys, expository articles, and research.

Major Contents: Theoretic methods on numerical analysis, approximation theory, functional analysis and operator optimization, control and systems theory. Also problems and models in the natural, physical, engineering, and decision sciences.

402 NUMERICAL HEAT TRANSFER

Hemisphere Publishing Corporation
1010 Vermont Avenue, N.W.
Washington, DC 20005

Editor: Professor W.J. Minkowycz, University of Illinois at Chicago, Department of Mechanical Engineering, P.O. Box 4348, Chicago, IL 60680
Phone: 312-996-3467
ISSN: 0149-5720 **Circulation:** 1,000 **Frequency:** BM
Style Sheet: Yes
Revised Theses or Diss.: Yes **Manuscript Copies:** 3
Review Period: 1-3 months (PR) **Lagtime:** 4 months

Types of Articles: Contribution and invited papers, technical notes, letters to the editor, numerical research, and book reviews by invitation.

Major Contents: Methodology and the results of numerical solutions of problems in heat and mass transfer, and related fluid flow.
Miscellaneous: An International Journal of Computation and Methodology.

403 **OECD OBSERVER, THE**

Organization for Economic Cooperation and Development
2 Rue Andre-Pascal
75775 Paris Cedex 16
France

Editor: Jane Bussiere
Phone: Paris 524-8200
ISSN: 0304-3398 **Circulation:** 25,000 **Frequency:** BM
Style Sheet: N/A
Revised Theses or Diss.: N/A **Manuscript Copies:** N/A
Review Period: N/A **Lagtime:** N/A
Types of Articles: Essays and research.
Major Contents: Covers the most important problems dealt with by member countries the OECD: energy economics and trends; interdependence; technological, environmental and social issues.
Miscellaneous: North American requests for information will be handled by the OECD Publications and Information Center, Suite 1207, 1750 Pennsylvania Avenue, N.W., Washington, DC 20006-4582, Phone: 202-724-1857.

404 **OPEC BULLETIN**

Public Information Department
Obere Donaustrasse 93
A-1020 Vienna
Austria

Editor: James Audu
Phone: 26-55-11, Ext. 280
ISSN: N/A **Circulation:** N/A **Frequency:** M
Style Sheet: Yes
Revised Theses or Diss.: No **Manuscript Copies:** 1
Review Period: 2 months **Lagtime:** 2 months
Types of Articles: Research; in-depth features on OPEC Member Countries and on the energy sectors of non-OPEC nations; texts of OPEC press releases; a news digest, market review and energy survey; details of international energy price movements; items from the OPEC News Agency.

Major Contents: Energy development issues; matters relating to: the oil market, new energy technology including alternative energy sources, and developmental economics.
Miscellaneous: This journal comprises a wide range of subject-matter, and is aimed at readers wishing to gain a broad insight into energy, development and related issues. Non-technical.

405 **OPEC REVIEW**

Pergamon Press, Ltd.
Headington Hill Hall
Oxford OX3 OBW
England

Editor: James Audu, Public Information Department, OPEC, Obere Donaustrasse 93, A-1020, Vienna, Austria
Phone: Vienna 26-55-11
ISSN: 0277-0180 **Circulation:** 2,000 **Frequency:** Q
Style Sheet: Yes
Revised Theses or Diss.: Yes **Manuscript Copies:** 2
Review Period: 3 months **Lagtime:** 1 month
Types of Articles: Research.
Major Contents: Topics relating to energy and development in the 13 OPEC Member Countries and other world nations including current energy issues, energy forecasts, alternative energy sources, north/south dialogue, and transfer of technology.

406 **OCEANUS**

Oceanus Subscriber Center
P.O. Box 6419
Syracuse, NY 13217

Editor: Paul R. Ryan, Oceanus, Woods Hole Oceanographic Institute, Woods Hole, MA 02543
Phone: 617-548-1400
ISSN: 0029-8182 **Circulation:** 15,000 **Frequency:** Q
Style Sheet: Chicago Manual of Style.
Revised Theses or Diss.: No **Manuscript Copies:** 2
Review Period: 3-6 months (PR) **Lagtime:** 6 months
Types of Articles: Research, reviews, and policy analysis.
Major Contents: General ocean topics: ocean energy, climate, deep-ocean drilling, the use of oceans to store waste, El Nino, deep sea hot springs and cold seeps, exclusive economic zones, etc.

407 **OFFSHORE - INTERNATIONAL JOURNAL OF OCEAN BUSINESS**

PennWell Publishing Company
1421 S. Sheridan Road, Box 1260
Tulsa, OK 74112

Editor: Eldon R. Ball, Managing Editor, 1200 Post Oak Boulevard, Suite 106, Houston, TX 77056
Phone: 713-621-9720
ISSN: 0030-0608 **Circulation:** 22,685 **Frequency:** M
Style Sheet: Yes
Revised Theses or Diss.: Yes **Manuscript Copies:** 1
Review Period: 1 week **Lagtime:** 1-3 months
Types of Articles: Technical, research and projects.
Major Contents: Offshore exploration and production, marine construction, marine transportation (boats and helicopters), submersibles, and remotely operated vehicles.
Miscellaneous: This journal serves the world-wide marine oil industry and Gulf Coast Region.

408 **OFFSHORE PETROLEUM INDUSTRY**

Barrows Company, Inc.
Box 1591, Grand Central Station
New York, NY 10163

Editor: M. Guerra
Phone: 212-772-1199
ISSN: N/A **Circulation:** N/A **Frequency:** N/A
Style Sheet: No
Revised Theses or Diss.: N/A **Manuscript Copies:** N/A
Review Period: N/A **Lagtime:** N/A
Major Contents: World oil and gas offshore operations in all countries.

409 **OIKOS**

Munksgaard International Booksellers and Publishers
35 Norre Sogade, DK-1370
Copenhagen K, Denmark

Editor: Per Brinck, Department of Animal Ecology, University of Lund, Ecology Building, S-223 62 Lund, Sweden
Phone: 46-46-109326
ISSN: 0030-1299 **Circulation:** 1,200 **Frequency:** BM
Style Sheet: See Miscellaneous
Revised Theses or Diss.: Yes **Manuscript Copies:** 3

Review Period: 1-3 months (PR) **Lagtime:** 8-12 months
Types of Articles: Research and reviews.
Major Contents: General ecology, theoretical ecology, experimental work, and the environmental impact of human activities.
Miscellaneous: Writing Scientific Papers in English.

410 **OIL AND GAS COMPACT BULLETIN**
Interstate Oil Compact Commission
Box 53127
Oklahoma City, OK 73152

Phone: 405-525-3556
ISSN: 0196-7177 **Circulation:** N/A **Frequency:** SA
Style Sheet: N/A
Revised Theses or Diss.: N/A **Manuscript Copies:** N/A
Review Period: N/A **Lagtime:** N/A
Types of Articles: Addresses and papers, committee reports.
Major Contents: The conservation of oil and gas through established state agencies.

411 **OIL AND GAS JOURNAL**
PennWell Publishing Company
1421 S. Sheridan Road, Box 1260
Tulsa, OK 74101

Editor: Gene T. Kinney
Phone: 918-835-3161
ISSN: 0030-1388 **Circulation:** 64,500 **Frequency:** W
Style Sheet: No
Revised Theses or Diss.: No **Manuscript Copies:** 1
Review Period: 1-2 weeks **Lagtime:** 1-2 months
Types of Articles: Technical, general articles, and projects.
Major Contents: Articles concerning the oil and gas industry which are of practical value to engineers, exploration managers, and operating economists.

412 **OPERATIONS RESEARCH**
Operations Research Society of America
Mount Royal and Guilford Avenues
Baltimore, MD 21202

Editor: Thomas L. Magnanti, Operations Research Center, Massachusetts Institute of Technology, Cambridge, MA 02139
Phone: 301-528-4146

ISSN: 0030-364X $ **Circulation:** /A$ **Frequency:** /year
Style Sheet: Writing a Technical Paper
Revised Theses or Diss.: N/A **Manuscript Copies:** 4
Review Period: 4 months (PR) **Lagtime:** N/A
Types of Articles: Research, technical notes, oriented applications, tutorials surveys, application papers, announcements, book reviews, reviews, letters to the editor, and forum.
Major Contents: Operations research methods data based operational science, and the practice or operations research, reporting underlying data based principles of operational science, observations of operating systems, contributions to the methods and models of operations research, case histories of applications, review articles, and disscussions of aspects of such subjects as the administrative environment, the history, policy, practice, future, or arenas of application of operations research, Complete studies that contain data, computer experiments, etc., and integrate the theory, methods, and applications are of particular interest.
Miscellaneous: The manuscript should be accompanied by a Copyright Transfer Agreement from the authors (or whoever holds the copyright) transferring the copyright to Operation Research Society of America. The form for this agreement is printed in every issue of the journal and is also available from the Editor and the Business Manager.

413 **OPTICAL AND QUANTUM ELECTRONICS**
Chapman and Hall, Ltd.
11 New Fetter Lane
London EC4 PEE
England

Editor: Dr. M.E. Barnett
ISSN: 0306-8919 **Circulation:** N/A **Frequency:** BM
Style Sheet: Yes
Revised Theses or Diss.: N/A **Manuscript Copies:** 2
Review Period: 2 weeks **Lagtime:** 6-8 months
Types of Articles: Research, reviews, letters, and communications.
Major Contents: Lasers, laser applications, nonlinear optics, optical modulations, light detection and imaging devices, semi-conductors, optoelectronic devices, thin film and integrated optics, optical communications, optical radar, velocimeters, holography, and space optics.
Miscellaneous: Formerly: OPTICS-ELECTRONICS, ISSN: 0030-4077.

414 **OPTICAL ENGINEERING**
Society of Photo-Optical Instrumentation Engineers
405 Fieldston Road, Box 10
Bellingham, WA 98227

Editor: Dr. H. John Caulfield
Phone: 206-676-3290
ISSN: 0091-3286 **Circulation:** 7,000 **Frequency:** BM
Style Sheet: Yes
Revised Theses or Diss.: N/A **Manuscript Copies:** 3
Review Period: 1-2 months (PR) **Lagtime:** 14 months
Types of Articles: Technical research, book reviews, full-length papers, and short communications.
Major Contents: New optical technology and previously discovered optical technology used in new ways including optical engineering, electro-optics, lasers, fiber optics, and optical systems.
Miscellaneous: Formerly: SPIE - THE INTERNATIONAL SOCIETY FOR OPTICAL ENGINEERING.

415 **PACIFIC OIL WORLD**
Petroleum Publishers, Inc.
222 S. Brea Boulevard, Box 129
Brea, CA 92621

Editor: H.H. Roberts, Editor-in-Chief
Phone: 213-691-1419
ISSN: 0008-1329 **Circulation:** 3,065 **Frequency:** M
Style Sheet: No
Revised Theses or Diss.: Yes **Manuscript Copies:** 1
Review Period: 1-2 weeks **Lagtime:** 1-4 weeks
Types of Articles: Research, technology, book reviews, project descriptions; research and technology reports, personality news.
Major Contents: Regional (Pacific) articles covering all aspects of drilling, refining, producing, pipelining, and manufacturing.
Miscellaneous: Formerly: CALIFORNIA WORLD OIL. Any theses or dissertations would need to conform to a journalistic style. This journal is written for the technical person in non-technical terms. All copies for the forthcoming issue are due by the 10th of the previous month.

416 **PARTICLE ACCELERATORS**
Gordon and Breach Science Publishing, Ltd.
1 Bedford Street
London WC2E 9TP
England

Editor: Francis Cole
Phone: 01-836-5125

ISSN: 0031-2460 **Circulation:** N/A **Frequency:** Q
Style Sheet: N/A
Revised Theses or Diss.: N/A **Manuscript Copies:** N/A
Review Period: N/A **Lagtime:** N/A
Types of Articles: Research, theoretical and experimental papers.
Major Contents: Particles orbit theory, high voltage techniques, magnetic design, vacuum system engineering, radiation shielding, accelerator instrumentation.

417 PARTICULATE SCIENCE AND TECHNOLOGY - AN INTERNATIONAL JOURNAL

Hemisphere Publishing Corporation
1010 Vermont Avenue, N.W.
Washington, DC 20005

Editor: J.K. Beddow, Coordinating Editor, Department of Chemical & Materials Engineering, University of Iowa, Iowa City, IA 52242
Phone: 319-353-3842
ISSN: 0272-6351 **Circulation:** 370 **Frequency:** 4/year
Style Sheet: Yes
Revised Theses or Diss.: Yes **Manuscript Copies:** 3
Review Period: N/A (PR) **Lagtime:** N/A
Types of Articles: Research, reviews , critical reviews, historical and biographical reports, technical notes, letters to the editor, book reviews, conference reports, and news items.
Major Contents: Topics include particle characterizations, particle sizing, morphological analysis, physical-chemical properties, instrumentation, particle formation and production, fluid particle separation, fluid particle interactions, particulate contamination, particle-particle separation, mixing, storage flow and handling, packing, agglomeration, aerosols, health hazards, materials safety, particulates processing, multiphase flow, fluidized beds, chemical engineering, mechanical engineering, pharmaceutical technology, ceramics, materials science, minerals processing, science and technology, powder metallurgy, soil, geology, image processing, pattern recognition, cement, pollution and pollution control, materials handling, and industrial hygiene.
Miscellaneous: An official publication of the Fine Particle Society.

418 PASSIVE SOLAR JOURNAL

American Solar Energy Society, Inc.
2030 17th Street
Boulder, CO 80302

Editor: Robert Jones

Phone: 303-443-3130
ISSN: 0277-6456 **Circulation:** N/A **Frequency:** IR
Style Sheet: Yes
Revised Theses or Diss.: Yes **Manuscript Copies:** 1
Review Period: N/A **Lagtime:** N/A
Types of Articles: Research, projects, and reviews.
Major Contents: Topics in passive systems, projects, architecture, funding, heating and cooling, hybrid technologies, strategies for architects, engineers, builders, homeowners and educators.
Miscellaneous: There is no PASSIVE SOLAR JOURNAL for 1984. Plans are indefinite for 1985.

419 **PATTERN RECOGNITION**
Pergamon Press
Maxwell House, Fairview Park
Elmsford, NY 10523

Editor: E.S. Gelsema, Managing Editor, Free University, Department of Medical Informatics, Vander Boechorststraat 7, 1081 BT Amsterdam, The Netherlands
Phone: 914-592-7700
ISSN: 0031-3203 **Circulation:** 1,500 **Frequency:** BM
Style Sheet: Yes
Revised Theses or Diss.: N/A **Manuscript Copies:** 3
Review Period: 1 month (PR) **Lagtime:** 6 months
Types of Articles: Research, reviews, pedagogical papers, symposia, and book reviews.
Major Contents: High-energy physics, target recognition, biological taxonomy, meteorology, space science, oceanography, character recognition, optical instrumentation, industrial applications and neuron physiology.

420 **PENNSYLVANIA GEOLOGY**
Bureau of Topographic and Geologic Survey
Department of Environmental Resources
P.O. Box 2357
Harrisburg, PA 17120

Editor: Dr. Arthur A. Socolow
Phone: 717-787-2169
ISSN: 0048-3214 **Circulation:** 7,000 **Frequency:** BM
Style Sheet: Yes
Revised Theses or Diss.: No **Manuscript Copies:** 1

Review Period: 2 weeks **Lagtime:** 2 months
Types of Articles: Research, projects, and new publications reviews.
Major Contents: Any subject dealing with geology, economic mineral and ground water resources which would be of interest to the general public of Pennsylvania.

421 **PETROLEUM ASIA JOURNAL**

Technical Editorial Wing
H.T. Building, 57-B, Rajpur Road
Box No. 50
Dehradun 248001, India

Editor: Dev K. Pandhi
Phone: 313023-RLKO-IN
ISSN: N/A **Circulation:** 10,000 **Frequency:** Q
Style Sheet: Yes
Revised Theses or Diss.: Yes **Manuscript Copies:** 2
Review Period: 3 months (PR) **Lagtime:** 3 months
Types of Articles: Research, technical and semi-technical papers, and world reviews in all disciplines of energy.
Major Contents: Research development in the energy sector of various countries, latest trends in the energy scene, analytical essays on the world oil scene, oil/gas forecasts for the world (in particular Asia), economics of the energy sector, oil, and gas hydrocarbon processing, and geoscientific perspectives.
Miscellaneous: Considered as the "foremost journal on oil and energy in India." Journal frequently devotes an entire issue to a selected topic. U.S. enquiries may be made at (212) 592-4913.

422 **PETROLEUM ECONOMIST**

Petroleum Press Bureau, Ltd.
107 Charterhouse Street
London EC1M 6AA
England

Editor: Bryan Cooper
Phone: London 01-251-3501
ISSN: 0306-395X **Circulation:** 7,000 **Frequency:** M
Style Sheet: Yes
Revised Theses or Diss.: No **Manuscript Copies:** 1
Review Period: 1 month **Lagtime:** 1-3 months
Types of Articles: Research, projects, and reviews.
Major Contents: Economic and/or financial aspects of energy developments, petroleum in particular.

Miscellaneous: Unsolicited articles are rare.

423 PETROLEUM ENGINEER INTERNATIONAL

Energy Publications
P.O. Box 1589
Dallas, TX 75221

Editor: W.B. Bleakley
Phone: 214-691-3911
ISSN: 0031-6466 **Circulation:** 42,000 **Frequency:** M
Style Sheet: No
Revised Theses or Diss.: No **Manuscript Copies:** 1
Review Period: 1 month **Lagtime:** 2-3 months
Types of Articles: Operations, case histories, and projects.
Major Contents: Drilling and production of oil and gas wells.
Miscellaneous: Need photo and author's biography.

424 PETROLEUM INDEPENDENT

Independent Petroleum Association of America
1101 16th Street, N.W.
Washington, DC 20036

Editor: Joe Taylor
Phone: 202-857-4775
ISSN: 0747-2528 **Circulation:** 14,000 **Frequency:** BM
Style Sheet: Yes
Revised Theses or Diss.: Yes **Manuscript Copies:** 1
Review Period: 1 month (PR) **Lagtime:** 1-2 months
Types of Articles: Research, projects, and reviews.
Major Contents: Topics in domestic petroleum exploration and production of interest to independent oil and gas producers.
Miscellaneous: An official publication of the Independent Petroleum Association of America.

425 PETROLEUM NEWS: ASIA'S ENERGY JOURNAL

Petroleum News Southeast Asia, Ltd.
146 Prince Edward Road, 6th Floor
Kowloon, Hong Kong

Editor: Donald Speakman, Managing Editor
Phone: 3-805-294-5
ISSN: 0250-7765 **Circulation:** 8,000 **Frequency:** M
Style Sheet: Yes
Revised Theses or Diss.: No **Manuscript Copies:** 1

Review Period: 1 month **Lagtime:** 1 month
Types of Articles: Research and projects.
Major Contents: Topics on energy, (oil and gas, coal, alternative energy, and nuclear) that have an Asian angle, either describing a technique that is applicable to the Asian situation, or actually about Asia.
Miscellaneous: Articles between 200-1500 words in length. U.S. inquiries may be made at 202-726-6150.

426 **PETROLEUM REVIEW**
Institute of Petroleum
61 New Cavendish Street
London W1M 8AR
England

Editor: Geoffrey Mayhew
Phone: 01-636 1004
ISSN: 0020-3076 **Circulation:** 9,000 **Frequency:** M
Style Sheet: Yes
Revised Theses or Diss.: Yes **Manuscript Copies:** 1
Review Period: 1 month **Lagtime:** 2-6 months
Types of Articles: Technical, research, projects, and training.
Major Contents: Various aspects of the petroleum industry and related industries.

427 **PETROLEUM TIMES**
Business Press International, Ltd.
The Quadrant, Quadrant House
Room 1101
Sutton, Surrey SM2 5AS, England

Editor: Bart Collins
Phone: 01-661-3324
ISSN: 0141-4437 **Circulation:** 5,500 **Frequency:** M
Style Sheet: No
Revised Theses or Diss.: No **Manuscript Copies:** N/A
Review Period: N/A **Lagtime:** 1 month
Types of Articles: Projects and reviews.
Major Contents: Oil and gas exploration activities, oil and gas production development, refining, all forms of crude oil and product transportation, crude oil pricing, and product pricing especially in Europe.

428 **PHOTOGRAMMETRIC ENGINEERING AND REMOTE SENSING**
American Society of Photogrammetry
210 Little Falls Street
Falls Church, VA 22046

Editor: Dr. J. Case
Phone: 703-534-6617
ISSN: 0099-1112 **Circulation:** 10,000 **Frequency:** M
Style Sheet: Yes
Revised Theses or Diss.: Yes **Manuscript Copies:** 5
Review Period: 6 months (PR) **Lagtime:** 6 months
Types of Articles: Research and projects. General and specialized.
Major Contents: Image acquisition; aerial and terrestrial photography and other imaging systems; applications of imagery to photogrammetry mapping and close-range photogrammetry and remote sensing geology, forestry, hydrology and land use; and data acquisition.
Miscellaneous: An official publication of the American Society of Photogrammetry.

429 PHOTOGRAPHIC SCIENCE AND ENGINEERING

Society of Photographic Scientists
7003 Kilworth Lane
Springfield, VA 22151

Editor: Rodney Shaw
Phone: 703-642-9090
ISSN: 0031-8760 **Circulation:** 4,000 **Frequency:** BM
Style Sheet: Yes
Revised Theses or Diss.: N/A **Manuscript Copies:** 3
Review Period: 1 month (PR) **Lagtime:** 6 months
Types of Articles: Research.
Major Contents: Topics in photography and other related sciences including photosensitive systems, the design of photographic instruments, photographic optics and illuminants, the use of photography for scientific or engineering measurement, and photographic instrumentation and data recording.

430 PHOTOVOLTAICS: THE SOLAR ELECTRIC MAGAZINE

Fore Publishers, Inc.
P.O. Box 3269
Scottsdale, AZ 85257

Editor: Jo'lene Ralston
Phone: 602-994-1923
ISSN: 0732-4375 **Circulation:** 9,000 **Frequency:** M
Style Sheet: Yes
Revised Theses or Diss.: Yes **Manuscript Copies:** 1

Review Period: 1 month **Lagtime:** N/A
Types of Articles: Research and projects.
Major Contents: All photovoltaics: cell research, modules, arrays, systems, storage, and utility interface.
Miscellaneous: Projects should be oriented to the consumer, utilities, municipalities, or countries.

431 **PHYSICS AND CHEMISTRY OF MINERALS**

Springer-Verlag Publishers
175 Fifth Avenue
New York, NY 10010

Editor: C.T. Prewitt, State University of New York at Stony Brook, Department of Earth and Space Sciences, Stony Brook, NY 11794
Phone: 212-460-1500
ISSN: 0342-1791 **Circulation:** N/A **Frequency:** BM
Style Sheet: Yes
Revised Theses or Diss.: N/A **Manuscript Copies:** N/A
Review Period: 2-4 months (PR) **Lagtime:** N/A
Types of Articles: Articles and short communications.
Major Contents: Physical and chemical studies on minerals or solids related to minerals. Mineralogy and physics or chemistry with emphasis on applications of modern techniques or new theories and models to interpret atomic structures and physical or chemical properties of minerals.

432 **PHYSICS TODAY**

American Institute of Physics
335 E. 45th Street
New York, NY 10017

Editor: Harold L. Davis
Phone: 212-661-9404
ISSN: 0031-9228 **Circulation:** 71,500 **Frequency:** M
Style Sheet: Yes
Revised Theses or Diss.: No **Manuscript Copies:** 2
Review Period: 4 months (PR) **Lagtime:** 6-8 months
Types of Articles: Research of reviews.
Major Contents: Physics, including recent developments in experimental and theoretical physics.

433 **PIPELINE**

Oildom Publishing Company of Texas, Inc.
3314 Mercer Street
Houston, TX 77027

Editor: Oliver Klinger, Jr.
Phone: 713-622-0676
ISSN: N/A **Circulation:** 13,000 **Frequency:** 7/year
Style Sheet: Yes
Revised Theses or Diss.: No **Manuscript Copies:** 1
Review Period: 1 month **Lagtime:** 1-3 months
Types of Articles: Research, projects, and reviews.
Major Contents: Engineering, management, and operations of oil. Natural gas, products, and coal slurry pipelines; telecommunications/automation, and corrosion control.

434 **PIPELINE AND GAS JOURNAL**

Energy Publications, A Division of Harcourt Brace Jovanovich
800 Davis Building, Box 1589
Dallas, TX 75221

Editor: A. Dean Hale
Phone: 214-691-3911
ISSN: 0032-0188 **Circulation:** 28,000 **Frequency:** M
Style Sheet: New York Times.
Revised Theses or Diss.: Yes **Manuscript Copies:** 1
Review Period: 1-2 months **Lagtime:** 2-6 months
Types of Articles: Research and projects.
Major Contents: Petroleum and natural gas pipeline design, engineering, operation, maintenance, and construction. Gas distribution system design, engineering, construction, maintenance, and operation.

435 **PIPELINE AND UNDERGROUND UTILITIES CONSTRUCTION**

Oildom Publishing Company of Texas, Inc.
3314 Mercer Street
Houston, TX 77027

Editor: Oliver Klinger, Jr.
Phone: 713-622-0676
ISSN: 0032-0196 **Circulation:** 14,000 **Frequency:** M
Style Sheet: Yes
Revised Theses or Diss.: No **Manuscript Copies:** 1
Review Period: 1 month **Lagtime:** 3 months
Types of Articles: Construction articles, personal items, and proprojects.
Major Contents: Innovations in underground construction particularly the pipeline and utility projects, maintenance, management and safety features.

436 **PIPELINE DIGEST**
Universal News, Inc.
P.O. Box 55225
Houston, TX 77255

Editor: Judy Clark
Phone: 713-468-2626
ISSN: 0197-1506 **Circulation:** 9,390 **Frequency:** SA
Style Sheet: Yes
Revised Theses or Diss.: No **Manuscript Copies:** 1
Review Period: 2-3 months **Lagtime:** 2-3 months
Types of Articles: Research, projects, reviews, and personal proprofiles.
Major Contents: Pipeline construction, operation, and maintenance. New technology, equipment, and services related to pipeline industry - engineering, corrosion control, inspection, welding, coal slurry pipelines, cogeneration, water pipelines, government regulations, and industry economics.

437 **PIPELINE INDUSTRY**
Gulf Publishing Company
P.O. Box 2608
Houston, TX 77001

Editor: Don Lambert
Phone: 713-529-4301
ISSN: N/A **Circulation:** 27,099 **Frequency:** M
Style Sheet: Yes
Revised Theses or Diss.: N/A **Manuscript Copies:** 1
Review Period: 3-4 weeks **Lagtime:** N/A
Types of Articles: Editorials.
Major Contents: Design, construction, operation, maintenance, and management of crude oil, and products, pipe lines, gas transmission lines and gas distribution systems.

438 **PLANETARY ASSOCIATION FOR CLEAN ENERGY NEWSLETTER**
Planetary Association for Clean Energy, Inc.
100 Bronson Avneue, Suite 1001
Ottawa, Ontario
K1R 6G8, Canada

Editor: Dr. Monique Michrowski, 77 Metcalfe Street, Suite 212, Ottawa, Ontario K1P 5L6, Canada

Phone: 613-236-6265
ISSN: 0708-918X **Circulation:** 2,000 **Frequency:** Q
Style Sheet: No
Revised Theses or Diss.: Yes **Manuscript Copies:** 1
Review Period: 12 months **Lagtime:** 12 months
Types of Articles: Research, projects, reviews, and book reviews.
Major Contents: Advanced clean energy systems and the related scientific and technological concerns, new devices, technology transfer, economics, physics and chemistry basis for new technology. The term "energy" is viewed in its physical and metaphysical sense at scales ranging from the planetary scale to the individual.

439 PLANT ENGINEERING

Technical Publishing Company
1301 S. Grove, Box 1030
Barrington, IL 60010

Editor: Leo Spector
Phone: 312-381-1840
ISSN: 0032-082X **Circulation:** 96,000 **Frequency:** SM
Style Sheet: Technical Writing Handbook.
Revised Theses or Diss.: Yes **Manuscript Copies:** 1
Review Period: 1 month **Lagtime:** 5-7 months
Types of Articles: Technical.
Major Contents: Subjects of job related interest and value to plant engineers in industry including the application of microcomputers, microprocessors, and related software to plant engineering jobs.

440 PLASMA PHYSICS AND CONTROLLED FUSION

The Institute of Physics
Techno House, Redcliffe Way
Bristol BS1 6NX
United Kingdom

Editor: Dr. A. Gibson, Honorary Editor
Phone: Bristol 027-2297-481
ISSN: 0032-1028 **Circulation:** 1,400 **Frequency:** M
Style Sheet: Yes
Revised Theses or Diss.: No **Manuscript Copies:** 3
Review Period: 3 months (PR) **Lagtime:** 8 months
Types of Articles: Research and reviews.
Major Contents: Controlled fusion, plasma physics.

441 **POLYMER BULLETIN**

Springer-Verlag Publishers
175 Fifth Avenue
New York, NY 10010

Editor: H.J. Cantow, J.P. Kennedy, T. Saegusa
Phone: 212-460-1500
ISSN: 0170-0839 **Circulation:** N/A **Frequency:** SA
Style Sheet: Yes
Revised Theses or Diss.: No **Manuscript Copies:** 2
Review Period: 3 weeks (PR) **Lagtime:** 2 months
Types of Articles: Brief reports of current research results.
Major Contents: Polymer science, biopolymers, chemistry, physical chemistry, physics, and materials science.
Miscellaneous: Review articles and abstracts of papers to be published elsewhere are not acceptable.

442 **POPULAR SCIENCE**

Times Mirror Magazines, Inc.
380 Madison Avenue
New York, NY 10017

Editor: C.P. Gilmore
Phone: 212-687-3000
ISSN: 0032-4647 **Circulation:** 1,800,00 **Frequency:** M
Style Sheet: Yes
Revised Theses or Diss.: No **Manuscript Copies:** 1
Review Period: 1-2 months **Lagtime:** 3 months
Types of Articles: Research, projects, and reviews.
Major Contents: New developments in science and technology in the fields of aerospace, physics, energy, transportation, and communication.
Miscellaneous: Articles must be targeted to a non-technical but knowledgeable readership.

443 **POWER**

McGraw-Hill Publishing Company
1221 Avenue of the Americas
New York, NY 10020

Editor: Robert G. Schwieger, Editor-in-Chief
Phone: 212-997-1221
ISSN: 0032-5929 **Circulation:** 55,000 **Frequency:** M
Style Sheet: No

Revised Theses or Diss.: No **Manuscript Copies:** 1
Review Period: 2 weeks **Lagtime:** 3 months
Types of Articles: Research, projects, and reviews.
Major Contents: Focus on modern technology basic to the function of modernizing and applying energy and energy equipment including instrumentation and control, steam generation, energy management, pollution control, pumps, compressors, valves, piping, turbines and diesels, fuels and fuel handling.

444 POWER CONVERSION AND INTELLIGENT MOTION

Intertec Communications, Inc.
2472 Eastman Avenue, Building 34
Ventura, CA 93003

Editor: Myron Miller
Phone: 805-658-0933
ISSN: 0199-1884 **Circulation:** 34,000 **Frequency:** M
Style Sheet: No
Revised Theses or Diss.: No **Manuscript Copies:** 1
Review Period: 1-2 months **Lagtime:** 3-4 months
Types of Articles: Research, technology innovations, and book reviews.
Major Contents: Power conversion and power supply design.
Miscellaneous: Of interest to electrical engineers and those in power business. Formerly: POWER CONVERSION INTERNATIONAL.

445 POWER ENGINEERING

Technical Publishing Company
1301 S. Grove, Box 1030
Barrington, IL 60010

Editor: John Papamarcos
Phone: 312-381-1840
ISSN: 0032-5961 **Circulation:** 12,000 **Frequency:** M
Style Sheet: No
Revised Theses or Diss.: No **Manuscript Copies:** 1
Review Period: 2 weeks **Lagtime:** 4 months
Types of Articles: Projects, reviews, and applications of engineers in the field.
Major Contents: All topics related to power plants planning, design, construction, operation, maintenance, economics, and costs. All types of energy conversion useful to industrial or utility power plants.
Miscellaneous: Abstract preferred for initial review.

446 **POWER LINE**

Environmental Action Foundation
724 Dupont Circle Building
Washington, DC 20036

Editor: Scott Ridley
Phone: 202-655-4000
ISSN: N/A **Circulation:** 2,300 **Frequency:** M
Style Sheet: Yes
Revised Theses or Diss.: Yes **Manuscript Copies:** 1
Review Period: 1 month **Lagtime:** 2-3 months
Types of Articles: Research, analysis, opinion, and state and local developments.
Major Contents: Topics in electric, and to a lesser extent gas, utilities including rates, rate hikes, generation economics, consumer and environmental policies, nuclear power, industry finance, conservation and renewables.
Miscellaneous: Contact the editor before submitting unsolicited manuscripts.

447 **PRECAMBRIAN RESEARCH**

Elsevier Science Publishers B.V.
P.O. Box 211
1000 AE Amsterdam
The Netherlands

Editor: B. Nagy, Editorial Office, Precambrian Research, P.O. Box 1930, 1000 AH Amsterdam, The Netherlands
Phone: 020-5803-911
ISSN: 0301-9268 **Circulation:** N/A **Frequency:** Q
Style Sheet: Yes
Revised Theses or Diss.: N/A **Manuscript Copies:** 3
Review Period: Varies **Lagtime:** Varies
Types of Articles: Research.
Major Contents: Precambrian research covers, but is not restricted to subjects such as: chemical, prebiological, biochemical and cosmochemical evolution, the origin of life, the evolution of the atmosphere and oceans, the fossil record, sedimentology, glaciation, stratigraphy, geochronology and isotope geochemistry; economic geology, geophysical aspects of the early Earth and Precambrian terrains; nature and evolution of the Precambrian lithosphere and mantle, magmatism, metamorphism, orogenesis, structural geology and plate tectonics. Studies on all aspects of the early stages of history and evolution of the earth and its planetary neighbours.

448 **PROCESS ENGINEERING**
Morgan-Grampian PLC
30 Calderwood Street
Woolwich, London SE18 6QH
England

Editor: C.A. Tayler
Phone: 01-855-7777
ISSN: 0370-1859 **Circulation:** 21,000 **Frequency:** M
Style Sheet: No
Revised Theses or Diss.: No **Manuscript Copies:** 1
Review Period: 1-2 months **Lagtime:** 1-2 months
Types of Articles: Reviews and technical articles with practical industry application.
Major Contents: All aspects of energy conservation in the "process" industries—chemical, food, iron and steel, cement, papers, drink and tobacco, metal processing, brick, pottery, and petrochemicals. Also process technology, process control and instrumentation, project management, and plant construction.

449 **PROCESS INDUSTRIES CANADA**
Southam Business Publications, Ltd.
1450 Don Mills Road
Don Mills
Ontario M3B 2X7, Canada

Editor: Arthur Kendrick
Phone: 416-445-6641
ISSN: 0008-3186 **Circulation:** 23,000 **Frequency:** BM
Style Sheet: Canadian Press Style Book.
Revised Theses or Diss.: No **Manuscript Copies:** 1
Review Period: N/A **Lagtime:** 2 months
Types of Articles: Applications, case histories, and articles.
Major Contents: Energy management in chemical processing industries. Areas include technology, equipment, business, legislative, and regulatory matters concerning the chemical processing industries.

450 **PROGRESS IN CRYSTAL GROWTH AND CHARACTERIZATION**
Pergamon Press
Maxwell House, Fairview Park
Elmsford, NY 10523

Editor: B.R. Pamplin
Phone: 914-592-7700
ISSN: 0146-3535 **Circulation:** 1,200 **Frequency:** 8/year
Style Sheet: N/A
Revised Theses or Diss.: N/A **Manuscript Copies:** N/A
Review Period: Varies (PR) **Lagtime:** 3 months
Types of Articles: Reviews, conference reports, state-of-the-art reports, lists of upcoming meetings on crystal growth and characterization.
Major Contents: Such aspects of crystal growth and characterization as: crystals of semi-conductors and electronic materials, oxides, synthetic minerals, magnetic and optical crystals, organic crystals and metals, eutectics, thin film, the theory of crystal growth, and methods of assessing crystal growth and purity.

451 PROGRESS IN ENERGY AND COMBUSTION SCIENCE

Pergamon Press
Maxwell House, Fairview Park
Elmsford, NY 10523

Editor: N.A. Chigier, Department of Mechanical Engineering, Carnegie-Mellon University, Schenly Park, Pittsburgh, PA 15213
Phone: 914-592-7700
ISSN: 0360-1285 **Circulation:** 1,200 **Frequency:** Q
Style Sheet: Yes
Revised Theses or Diss.: N/A **Manuscript Copies:** 3
Review Period: N/A **Lagtime:** 9 months
Types of Articles: Research and reviews.
Major Contents: Combustion, flames, fuel science, technology, and energy studies.

452 PROGRESS IN NUCLEAR ENERGY

Pergamon Press
Maxwell House, Fairview Park
Elmsford, NY 10523

Editor: Professor L.E. Weaver, 108 Ramsay Hall, Auburn University, Auburn, AL 36849
Phone: 914-592-7700
ISSN: 0149-1970 **Circulation:** N/A **Frequency:** BM
Style Sheet: Yes
Revised Theses or Diss.: Yes **Manuscript Copies:** 2
Review Period: 1 month (PR) **Lagtime:** 6 months
Types of Articles: Research and reviews.

Major Contents: Nuclear power, nuclear safety, nuclear fuel reprocessing, fusion, and reactor physics.
Miscellaneous: Professor M.M.R. Williams, Editor (United Kingdom and rest of the world).

453 **PROPANE/CANADA**
Sanford Evans Publishing, Ltd.
1077 James Street, Box 6900
Winnipeg, Manitoba
R3C 3B1, Canada

Editor: Scott Jeffrey, 205 9th Avenue, S.E., Suite 503, Calgary, Alberta T2G OR3, Canada
Phone: 403-265-4750
ISSN: 0033-1260 **Circulation:** 5,500 **Frequency:** BM
Style Sheet: Yes
Revised Theses or Diss.: Yes **Manuscript Copies:** 2
Review Period: 2 months **Lagtime:** 2 months
Types of Articles: Research, projects, and product and literature reviews.
Major Contents: Miscible flood, petrochemicals from LPG feedstock, LP Gas carburetion, NGV carburetion, industry events, conventions, and government regulations. General and specialized articles deal with all aspects of the North American, LP Gas industry, including propane, butane, and pentanes plus.
Miscellaneous: Edited for the North American, especially the Canadian, LP Gas industry, and concentrates all efforts on the promotion and well-being of this industry.

454 **PUBLIC POWER**
American Public Power Association
2301 M Street, N.W.
Washington, DC 20037

Editor: Lori Woehrle, Assistant Editor
Phone: 202-775-8300
ISSN: 0033-3654 **Circulation:** 17,000 **Frequency:** BM
Style Sheet: Yes
Revised Theses or Diss.: No **Manuscript Copies:** 1
Review Period: 1 month **Lagtime:** 3-6 months
Types of Articles: Research, projects, and reviews.
Major Contents: All aspects of electric systems including conservation, distribution, energy services, environmental protection, financing, load management, new public power systems, rates, and service.

Miscellaneous: An official publication of the American Public Power Association.

455 PUBLIC UTILITIES FORTNIGHTLY

Public Utilities Reports, Inc.
1700 N. Moore Street, Suite 2100
Rosslyn Center Bldg.
Arlington, VA 22209

Editor: Lucien E. Smartt
Phone: 703-243-7000
ISSN: 0033-3808 **Circulation:** 8,000 **Frequency:** BW
Style Sheet: Yes
Revised Theses or Diss.: N/A **Manuscript Copies:** 2
Review Period: 4-6 months **Lagtime:** 2-9 months
Types of Articles: Policy, economic, business and management, articles, editorials, calendar, comments, and speeches.
Major Contents: Primarily devoted to economic, regulatory legislative, trade, management, marketing or environmental events/possibilities which could impact public utilities.
Miscellaneous: Articles limited to 20 typed double-spaced pages.

456 PURE AND APPLIED GEOPHYSICS (PAGEOPH)

Birkhauser Boston, Inc.
380 Green Street
Cambridge, MA 02139

Editor: Renata Dmowska, Executive Editor, Harvard University, Division of Applied Science, 29 Oxford Street, Pierce Hall, Cambridge, MA 02138
Phone: 617-495-3452
ISSN: 0033-4553 **Circulation:** 600 **Frequency:** BM
Style Sheet: No
Revised Theses or Diss.: Yes **Manuscript Copies:** 3
Review Period: 1 month (PR) **Lagtime:** 4-6 months
Types of Articles: Research and reviews.
Major Contents: All aspects of geophysics (seismology, magnetism, tectonophysics heat studies, etc.; ocean and atmospheric studies.
Miscellaneous: For orders outside the United States and Canada write to: Birkhauser Verlag, P.O. Box 133, CH-4010 Basel, Switzerland. Pageoph is a continuation of "Geofisica pura e applicata."

457 **RE MAGAZINE**

National Rural Electric Cooperative Association
1800 Massachusetts Avenue, N.W.
Washington, DC 20036

Editor: Frank Gallant, Managing Editor
Phone: 202-857-9500
ISSN: N/A **Circulation:** 25,000 **Frequency:** M
Style Sheet: *See Miscellaneous
Revised Theses or Diss.: No **Manuscript Copies:** 1
Review Period: 1 month **Lagtime:** 2-3 months
Types of Articles: Articles and news of interest to managers and directors of rural electric cooperatives.
Major Contents: Special activities of rural electric cooperatives, electric utility management, new trends in the energy field, and rural development of electric co-ops.
Miscellaneous: Formerly: RURAL ELECTRIFICATION. *New York Times Manual of Style and Usage.

458 **RADIATION PHYSICS AND CHEMISTRY**

Pergamon Press
Maxwell House, Fairview Park
Elmsford, NY 10523

Editor: Professor A. Charlesby, Editor-in-Chief, Silverspring, Eagle Lane, Watchfield, Swindon, Wiltshire SN6 8LA, United Kingdom
Phone: 914-592-7700
ISSN: 0146-5724 **Circulation:** 1,500 **Frequency:** M
Revised Theses or Diss.: N/A **Manuscript Copies:** N/A
Review Period: N/A **Lagtime:** N/A
Types of Articles: Research.
Major Contents: Radiation physics and chemistry, including radiation reactions, processes, and industrial applications.
Miscellaneous: Authors are requested to send manuscripts to the Regional Editors: Dr. A. Singh, American Editor, Medical Biophysics Branch, Whiteshell Nuclear Research Establishment, Atomic Energy of Canada, Ltd., Pinawa, Manitoba, Canada, ROE 1L0; Professor R. H. Schuler, Director, Radiation Laboratory, Notre Dame, IN 46556

459 **RADIOACTIVE WASTE MANAGEMENT AND THE NUCLEAR FUEL CYCLE**

Harwood Academic Publishers
50 W. 23rd Street
New York, NY 10010

Editor: A.M. Platt, Battelle Pacific Northwest Laboratories, Richland, WA 99352
Phone: 509-375-2273
ISSN: 0142-2405 **Circulation:** N/A **Frequency:** Q
Style Sheet: Yes
Revised Theses or Diss.: Yes **Manuscript Copies:** 3
Review Period: 1 months (PR) **Lagtime:** 3 months
Types of Articles: Research and development.
Major Contents: Specific research may relate to the disposal and waste management processes of radioactive waste treatment, conditioning, transportation, interim storage, disposal site selection, repository design, ultimate disposal, management strategies, cost benefit analyses, and reactor operations impacting waste management.
Miscellaneous: Manuscripts should be submitted double spaced with wide margins (3 cm) on good quality paper. Abstract of 100-150 words.

460 RENEWABLE ENERGY NEWS

Solar Vision, Inc.
7 Church Hill
Harrisville, NH 03450

Editor: William D'Alessandro
Phone: 603-827-3347
ISSN: 0714-8615 **Circulation:** 16,000 **Frequency:** M
Style Sheet: Yes
Revised Theses or Diss.: Yes **Manuscript Copies:** 1
Review Period: 2 weeks **Lagtime:** 4 months
Types of Articles: Press releases, project summaries, annual institutional reports, and research.
Major Contents: Renewable sources of energy, energy conservation and management, marketing, development, investment, solar, wind, small scale hydro, geothermal, low energy buildings, heat pumps, and energy storage.
Miscellaneous: Incorporated SOLAR TIMES (1982) and NEW ROOTS (1983).

461 RESEARCH AND DEVELOPMENT MANAGEMENT DIGEST

Lomond Publications, Inc.
P.O. Box 88
Mt. Airy, MD 21771

Editor: Lowell H. Hattery, Ph.D.
Phone: 301-829-1496

ISSN: 0361-753X **Circulation:** N/A **Frequency:** M
Style Sheet: No
Revised Theses or Diss.: No **Manuscript Copies:** 1
Review Period: 1 month **Lagtime:** 2 months
Types of Articles: Research, case experience reports, announcements of meetings and development opportunities, literature reviews, interdisciplinary reports, and technological forecasting.
Major Contents: Computer and information science, planning theory, organization theory, behavioral science, human relations, industrial engineering, and laboratory management.
Miscellaneous: A publication for brief announcements and informational articles which would be of interest to R&D managers.

462 **RESEARCH AND DEVELOPMENT: INTERNATIONAL JOURNAL OF RESEARCH, DEVELOPMENT, AND QUALITY ASSURANCE**
Technical Publishing Company
P.O. Box 1030
Barrington, IL 60010

Editor: Robert R. Jones
Phone: 312-381-1840
ISSN: 0160-4074 **Circulation:** 107,103 **Frequency:** M
Style Sheet: Yes
Revised Theses or Diss.: Yes **Manuscript Copies:** 1
Review Period: 3 months (PR) **Lagtime:** 3 months
Types of Articles: Applied research and development interpretive news and articles of interest and potential use to persons engaged in research, development, and quality assurance "state-of-the-art" articles.
Major Contents: Industrial research, development, and quality assurance. Analytical chemistry; applied spectroscopy; chemical processes; computer hardware and software; biotechnology; fossil fuel; fission; fusion, solar energy; and nuclear power.
Miscellaneous: Formerly: INDUSTRIAL RESEARCH AND DEVELOPMENT

463 **RESOURCES**
Resources for the Future, Inc.
1616 P Street, N.W.
Washington, DC 20036

Editor: Kent Price
Phone: 202-328-5025/5026

ISSN: 0048-7376 **Circulation:** 20,000 **Frequency:** Q
Style Sheet: Chicago Manual of Style.
Revised Theses or Diss.: No **Manuscript Copies:** 1
Review Period: 1 month **Lagtime:** 3 months
Types of Articles: Editorials, research, news, and reviews.
Major Contents: Research, policy, and economics relating to the many and varied aspects of the development, conservation and use of natural resources, and environmental quality.
Miscellaneous: This publication does not ordinarily accept outside articles but does occasionally accept outstanding submissions.

464 **RESOURCES AND CONSERVATION**

Elsevier Science Publishers B.V.
P.O. Box 330
1000 AH Amsterdam
The Netherlands

Editor: Dr. H. Alter, Manager, Resources Policy Department, Chamber of Commerce of the United States, 1615 H. Street, N.W., Washington, DC 20062
Phone: 202-463-5531
ISSN: 0166-3097 **Circulation:** 400 **Frequency:** Q
Style Sheet: Yes
Revised Theses or Diss.: Yes **Manuscript Copies:** 3
Review Period: 1-2 months (PR) **Lagtime:** 2-3 months
Types of Articles: Investigations, analyses, reviews, and scientific technological or institutional methodology.
Major Contents: The journal is dedicated to detailed and comprehensive investigations, analyses, and appropriate reviews of the interdisciplinary aspects of renewable and nonrenewable resource management. Particular emphasis is given to resource conservation. Topics include: new technologies and management methods for reducing consumption and the substitution of nonrenewable resources; recovery of materials and production of energy from established and new sources such as domestic, commercial and industrial wastes; management of natural resources, including implementation of public policies for conservation; management of processes and product design; fabrication and production for efficient utilization of materials and energy.
Miscellaneous: Formerly: RESOURCE RECOVERY AND CONSERVATION, ISSN: 0304-3967.

465 **RESOURCES AND ENERGY**

Elsevier Science Publishers B.V.
P.O. Box 211
1000 AE Amsterdam
The Netherlands

Editor: George S. Tolley, Editorial Offices, 104 S. Michigan Avenue, Suite 717, Chicago, IL 60603
Phone: 312-726-1541
ISSN: 0165-0572 **Circulation:** N/A **Frequency:** Q
Style Sheet: Yes
Revised Theses or Diss.: N/A **Manuscript Copies:** 3
Review Period: 3-6 months (PR) **Lagtime:** 6 months
Types of Articles: Theoretical models, empirical studies, and applications of techniques.
Major Contents: An international journal which provides a high level forum on use of the earth's resources for the benefit of mankind. Applications of techniques include econometrics, systems analysis, growth models and energy, environmental and resource analysis.

466 **RESOURCES POLICY**

Butterworth Scientific, Ltd.
Box 63, Westbury House, Bury Street
Guildford GU2 5BH
England

Editor: L. Driscoll
Phone: Guildford 0483-31261
ISSN: 0301-4207 **Circulation:** 3,500 **Frequency:** Q
Style Sheet: Yes
Revised Theses or Diss.: Yes **Manuscript Copies:** 3
Review Period: 1-3 months (PR) **Lagtime:** 6-10 months
Types of Articles: Research, review/overview papers, discussions, rejoinders, letters, conference reports, and book reviews.
Major Contents: Economic, political, social, and planning aspects of nonfuel minerals supply and demand such as: resource/reserve assessment, estimation, international policy, commodity agreements, trade effects, strategic minerals, mining agreements, substitution/recycling/conservation aspects, environmental aspects, world/national modelling, mining investment, taxation; and seabed minerals.

467 **REVIEW OF NEW ENERGY TECHNOLOGY**

P.O. Box 609
Littleton, CO 80160

Editor: Charles O. Hook
Phone: 303-795-7487
ISSN: 0748-4011 **Circulation:** N/A **Frequency:** M
Style Sheet: No
Revised Theses or Diss.: Yes **Manuscript Copies:** 1
Review Period: 3 months (PR) **Lagtime:** 3 months
Types of Articles: Research, projects, reviews of recent patents, analysis, international developments, government activities, conference papers, and an occasional editorial.
Major Contents: International energy developments; new energy research and development; R&D in the field of unconventional energy i.e. energy minerals research, synthetic fuels and alternate energy such as coal gasification, geothermal, nuclear, enhanced oil recovery, MHD, oceans, oil shale, peat, tar sands, solar, wind, biomass; and legislation, corporate activities, and policy.
Miscellaneous: Entire papers or dissertations are rarely accepted. The journal reserves the right to edit and excerpt.

468 REVIEWS OF GEOPHYSICS AND SPACE PHYSICS

The American Geophysical Union
2000 Florida Avenue, N.W.
Washington, DC 20009

Editor: James R. Heirtzler, Woods Hole Oceanographic Institution, Woods Hole, MA 02543
Phone: 617-548-1400
ISSN: 0034-6853 **Circulation:** 2,500 **Frequency:** Q
Style Sheet: AGU Style Manual.
Revised Theses or Diss.: N/A **Manuscript Copies:** 3
Review Period: 1-2 months (PR) **Lagtime:** 9 months
Types of Articles: Full-length reviews and topical reviews.
Major Contents: General topics in geophysics, especially hydrology, tectonics, seismology, and oceanography.
Miscellaneous: Formerly: REVIEWS OF GEOPHYSICS.

469 ROBOTICS AND AUTOMATION

Acta Press
P.O. Box 25, Station G
Calgary, Alberta
T3A 2G1, Canada

Editor: S. Hsia
Phone: 403-270-3616

ISSN: N/A **Circulation:** N/A **Frequency:** 3/year
Style Sheet: No
Revised Theses or Diss.: Yes **Manuscript Copies:** 3
Review Period: 2 months (PR) **Lagtime:** N/A
Types of Articles: Research, book reviews, conferences, new publications, advertisements, and call for papers.
Major Contents: Modelling, simulation, kinematics, dynamics, control, robot-computer interface, software, languages, bionics and robots, teleoperators, sensors, vision, safety, supervision, industrial robots, expert systems, multirobot systems, and applications in all fields.

470 ROCK MECHANICS AND ROCK ENGINEERING
Springer Verlag Publishers
175 Fifth Avenue
New York, NY 10010

Editor: Professor K. Kovori, Federal Institute of Technology Zurich, ETH Honggerberg, 8093 Zurich, Switzerland
Phone: 212-460-1500
ISSN: 0723-2632 **Circulation:** 1,500 **Frequency:** Q
Style Sheet: Yes
Revised Theses or Diss.: N/A **Manuscript Copies:** 3
Review Period: 2 months (PR) **Lagtime:** 5-6 months
Types of Articles: Research, case histories, and projects.
Major Contents: Experimental and theoretical aspects of rock engineering.
Miscellaneous: Formerly: ROCK MECHANICS, ISSN: 0035-7448.

471 SCANDANAVIAN OIL-GAS MAGAZINE
Scandinavian Oil-Gas Publishing Company
P.O. Box 6865
St. Olavs Plass
0130 Oslo 1, Norway

Editor: Odd E. Pederson
Phone: 02-447270/447287
ISSN: 0332-5334 **Circulation:** 11,000 **Frequency:** BM
Style Sheet: No
Revised Theses or Diss.: Yes **Manuscript Copies:** 1
Review Period: 1-2 months **Lagtime:** 1-2 months
Types of Articles: Research, projects, reviews, interviews, exhibition reviews, and business profiles.
Major Contents: Offshore oil and gas exploration, reserves, production, marketing, energy policy, and offshore technology.

Miscellaneous: International in coverage.

472 **SCIENCE**

American Association for the Advancement of Science
1515 Massachusetts Avenue, N.W.
Washington, DC 20005

Editor: Daniel E. Koshland, Jr.
Phone: 202-467-4490
ISSN: 0036-8075 **Circulation:** 160,000 **Frequency:** W
Style Sheet: Yes
Revised Theses or Diss.: Yes **Manuscript Copies:** 3
Review Period: 1-2 months (PR) **Lagtime:** 3 weeks
Types of Articles: Research, projects, and reviews.
Major Contents: Topics in physical and life sciences relating to the advancement of science.
Miscellaneous: Articles are written by professionals for professionals in fields other than the one being discussed.

473 **SCIENCE DIGEST**

Hearst Magazines
888 7th Avenue
New York, NY 10106

Editor: Oliver S. Moore, III
Phone: 212-262-7990
ISSN: 0036-8296 **Circulation:** 530,000 **Frequency:** M
Style Sheet: Yes
Revised Theses or Diss.: Yes **Manuscript Copies:** 1
Review Period: 1-2 months **Lagtime:** 3 months
Types of Articles: News, features, and editorials.
Major Contents: All areas of scientific and technological development from innovation and research to policy.
Miscellaneous: This journal was decidedly written for the male audience; the editor states "Science Digest is a whole new kind of men's magazine," which attempts to demystify science." Contributors are frequently researchers who present the most complex subjects in understandable terms and style.

474 **SCIENCE OF THE TOTAL ENVIRONMENT, THE**

Elsevier Science Publishers B.V.
P.O. Box 211
1000 AE Amsterdam
The Netherlands

Editor: Dr. E.I. Hamilton, "Woodchurch" Crapstone, Yelverton, Devon, United Kingdom
Phone: 020-5803-911
ISSN: 0048-9697 **Circulation:** 500 **Frequency:** M
Style Sheet: Yes
Revised Theses or Diss.: Yes **Manuscript Copies:** 3
Review Period: 3 weeks (PR) **Lagtime:** 9 months
Types of Articles: Research, projects, reviews, technical notes, symposia and conference proceedings, and letters to the editor.
Major Contents: Changes in the environment caused by man's activities, and changes in the natural level and distribution of chemical elements and compounds which may effect the well-being of the living world with emphasis to applied biogeochemistry and medical environmental research. The subjects covered include: (a) application of techniques and methods of chemistry and biochemistry to environmental problems; (b) pollution of the air, water, soil and various aspects of human nutrition; (c) environmental medicine, when the effect of abnormalities in the level and distribution of chemical elements and compounds are given prominence; (d) the use of interdisciplinary methods in studies of the environment; (e) environmental planning and policy.
Miscellaneous: English is the preferred language, though occasional articles in French and German are accepted.

475 **SCIENTIA**

Via Guastalla 9
20122 Milano, Italy

Editor: N. Bonetti
Phone: Italy 02-780-669
ISSN: 0036-8687 **Circulation:** 11,800 **Frequency:** 3/year
Style Sheet: N/A
Revised Theses or Diss.: N/A **Manuscript Copies:** N/A
Review Period: N/A **Lagtime:** N/A
Types of Articles: Research, reports, book reviews, and books received.
Major Contents: All fields of scientific knowledge from the history of science to mathematics; astronomy; physics; chemistry; biology in all its various specializations; psychology; social sciences, etc.
Miscellaneous: SCIENTIA is a review with worldwide circulation since 1907. It aims at being a means of communication between specialists in various disciplines that can promote a synthetic vision and a critical examination of the problems of science. All articles are written at the editor's request.

476 **SCIENTIFIC AMERICAN**
Scientific American, Inc.
415 Madison Avenue
New York, NY 10017

Editor: Dennis Flanagan
Phone: 212-754-0550
ISSN: 0036-8733 **Circulation:** 715,000 **Frequency:** M
Style Sheet: Yes
Revised Theses or Diss.: Yes **Manuscript Copies:** 2
Review Period: 1-2 months (PR) **Lagtime:** 3-5 months
Types of Articles: Research, projects, and reviews.
Major Contents: Multidisciplinary articles that interpret scientific and technological advances and theories. Also articles about new R&D.
Miscellaneous: Scientific topics are written for an educated but non-technical public.

477 **SCIQUEST**
American Chemical Society
1155 16th Street, N.W.
Washington, DC 20036

Editor: Patricia A. Morgan, Managing Editor
Phone: 202-872-4600
ISSN: N/A **Circulation:** 38,390 **Frequency:** 10/year
Style Sheet: Yes
Revised Theses or Diss.: No **Manuscript Copies:** 2
Review Period: 1-2 months (PR) **Lagtime:** 3 months
Types of Articles: Research, projects, reviews, and non-technical articles.
Major Contents: Atomic and molecular activity with specific emphasis on application.
Miscellaneous: Articles in contemporary science presented to entertain as well as inform a primarily non-technical audience.

478 **SCRIPTA METALLURGICA**
Pergamon Press
Maxwell House, Fairview Park
Elmsford, NY 10523

Editor: J.P. Hirth
Phone: 914-592-7700
ISSN: 0036-9748 **Circulation:** 1,600 **Frequency:** M
Style Sheet: Yes

Revised Theses or Diss.: N/A **Manuscript Copies:** 2
Review Period: 1-2 months (PR) **Lagtime:** 3 months
Types of Articles: Short reports of new work and discussions of previously published work.
Major Contents: Topics on the physical properties of such materials as: metals and alloys, ceramics, solid polymers, and glasses. Papers linking the microscopic to the macroscopic properties of materials are preferred.

479 SEMICONDUCTORS AND INSULATORS

Gordon and Breach Science Publishers, Inc.
1 Park Avenue
New York, NY 10010

Editor: R. Smoluchowski
Phone: 212-206-8900
ISSN: 0039-5991 **Circulation:** N/A **Frequency:** Q
Style Sheet: Yes
Revised Theses or Diss.: N/A **Manuscript Copies:** 2
Review Period: 1 month **Lagtime:** 3 months
Types of Articles: Research, reviews, letters, and short communications.
Major Contents: Atomic and molecular level of electronic and structural properties of semi-conductors, ionic, covalent, van der Waals, hydrogen-bonded crystalline, noncrystalline liquids and solids.

480 SEPARATION SCIENCE AND TECHNOLOGY

Marcel Dekker, Inc.
270 Madison Avenue
New York, NY 10016

Editor: Dr. J. Calvin Giddings, P.O. Box 11305, Church Street Station, New York, NY 10049
Phone: 212-696-9000
ISSN: 0149-6395 **Circulation:** 938 **Frequency:** 15/year
Style Sheet: Yes
Revised Theses or Diss.: N/A **Manuscript Copies:** 3
Review Period: 2 months **Lagtime:** 10 months
Types of Articles: Research, communications, notes, and reviews.
Major Contents: Topics concerning separation which develop new concepts and techniques for the field.
Miscellaneous: Formerly: SEPARATION SCIENCE, ISSN: 0037-2366.

481 SIERRA

Sierra Club
530 Bush Street
San Francisco, CA 94108

Editor: James Keough, Jonathon King, Associate Editor
Phone: 415-981-8634
ISSN: 0161-7362 **Circulation:** 335,000 **Frequency:** BM
Style Sheet: Yes
Revised Theses or Diss.: No **Manuscript Copies:** 1
Review Period: 1-2 months **Lagtime:** 3-6 months
Types of Articles: Articles, photographs and art on environmental topics, issues and places currently important to environmentalists.
Major Contents: Preservation of wilderness, national parks, natural areas, social responsibility, holistic thinking, controlled growth, sound urban planning, and energy conservation.
Miscellaneous: Authors are encouraged to submit queries or story ideas before writing an article.

482 SOCIETY OF PETROLEUM ENGINEERS JOURNAL

Society of Petroleum Engineers
P.O. Box 833836
Richardson, TX 75083

Editor: Jim McInnis
Phone: 214-669-3377
ISSN: 0197-7520 **Circulation:** 16,500 **Frequency:** BM
Style Sheet: Yes
Revised Theses or Diss.: No **Manuscript Copies:** 3
Review Period: 3-6 months (PR) **Lagtime:** 3-9 months
Types of Articles: Projects, reviews, reports on developments in the laboratory, trial field tests, and other theoretical or research findings.
Major Contents: Any phase of petroleum drilling, exploration, or production, including chemical systems and computer applications.

483 SOL NEWSLETTER

Solar Energy Society of Canada, Inc.
303-870 Cambridge Street
Manitoba MB R3M 3H5
Canada

Editor: Bill Eggertson
Phone: 613-236-4594
ISSN: N/A **Circulation:** 3,000 **Frequency:** BM
Style Sheet: No
Revised Theses or Diss.: Yes **Manuscript Copies:** 1
Review Period: N/A (PR) **Lagtime:** 2 months
Types of Articles: Research, projects, reviews, equipment and techno-

logical developments, and book reviews.
Major Contents: Energy conservation; sources and use of direct solar (active, passive, photovoltaic), and other renewable energy; emphasis on low energy buildings.

484 **SOLAR AND WIND TECHNOLOGY**
Pergamon Press
Maxwell House, Fairview Park
Elmsford, NY 10523

Editor: Dr. A.A.M. Sayigh, Energy Resources, OAPEC, 20501, Safat, Kuwait
Phone: 244-8200, Ext. 323
ISSN: N/A **Circulation:** N/A **Frequency:** Q
Style Sheet: No
Revised Theses or Diss.: Yes **Manuscript Copies:** 3
Review Period: 2 months **Lagtime:** 6 months
Types of Articles: Research, special projects, reviews, state-of-the-art articles, inventions, and book reviews.
Major Contents: Practical experience and technology assessment in solar and wind area.

485 **SOLAR CELLS**
Elsevier Sequoia S.A.
P.O. Box 851
CH-1001 Lausanne 1
Switzerland

Editor: Dr. L.L. Kazmerski, Solar Energy Research Institute, 1536 Cole Boulevard, Golden, CO 80401
Phone: 303-231-1115
ISSN: 0379-6787 **Circulation:** 600 **Frequency:** M
Style Sheet: Yes
Revised Theses or Diss.: Yes **Manuscript Copies:** 3
Review Period: 1-4 months (PR) **Lagtime:** 6-8 months
Types of Articles: Research, reviews, short communications, book reviews, announcements, reports on conferences, and news.
Major Contents: Improvement of device efficiency, reduction of cost, testing and applications of solar cells; physics and chemistry of solar cells including single-crystal, polycrystalline and amorphous homojunction-heterojunction and schottky devices; specific applications ranging from centralized power stations to remote power supplies, and from rural electrification to military uses; economic aspects covering cost forecasting; the implications of materials costs and resources; produc-

tion technology and reliability; social implications relating to the environment and the likely influence on rural communities; systems analysis of individual cells and complete generators such as flat plate collectors and concentrators.

486 **SOLAR ENERGY**

Pergamon Press
Maxwell House, Fairview Park
Elmsford, NY 10523

Editor: Dr. Peter E. Glaser, Arthur D. Little, Inc., Acorn Park, Cambridge, MA 02140
Phone: 617-864-5770
ISSN: 0038-092X **Circulation:** 13,000 **Frequency:** M
Style Sheet: AIP Style Manual.
Revised Theses or Diss.: Yes **Manuscript Copies:** 4
Review Period: 3-8 months (PR) **Lagtime:** 3-6 months
Types of Articles: Scientific or engineering papers describing original studies, technical notes, state-of-the-art reviews, and topical reports.
Major Contents: Any aspect of solar energy research, measurement, development or application: radiation and measurement, heating and cooling, photochemistry, photovoltaics, wind, solar thermal, biomass, and bioconversion.
Miscellaneous: An official journal of the International Solar Energy Society.

487 **SOLAR ENERGY INTELLIGENCE REPORT**

Business Publishers, Inc.
751 Pershing Drive
Silver Spring, MD 20910

Editor: Allan L. Frank
Phone: 301-587-6300
ISSN: 0148-4095 **Circulation:** 30 **Frequency:** W
Style Sheet: Yes
Revised Theses or Diss.: No **Manuscript Copies:** 1
Review Period: 1-2 months **Lagtime:** 1-2 months
Types of Articles: Projects, news, and reviews.
Major Contents: Practical solar, solar energy business and marketing reports, solar outlook, solar energy industry report, solar energy projects and policy.
Miscellaneous: Trade newsletter covers current activities, research and development in news brief format. Authors are not credited.

488 **SOLAR ENERGY MATERIALS**

Elsevier Science Publishers B.V.
P.O. Box 211
1000 AE Amsterdam
The Netherlands

Editor: C.M. Lampert, Lawrence Berkeley Laboratory, Materials and Molecular Resource Division, Cyclotron Road, Building 62, Berkeley, CA 94720
ISSN: 0165-1633 **Circulation:** N/A **Frequency:** BM
Style Sheet: Yes
Revised Theses or Diss.: No **Manuscript Copies:** 3
Review Period: 3 weeks **Lagtime:** 4-6 months
Types of Articles: Research.
Major Contents: Materials science aspects of photovoltaic, photothermal, and photochemical solar energy conversion.

489 **SOLAR ENGINEERING AND CONTRACTING**

Business News Publishing Company
Box 3600
Troy, MI 48007

Editor: Timothy Fausch
Phone: 313-362-3700
ISSN: 0731-2970 **Circulation:** 16,000 **Frequency:** BM
Style Sheet: No
Revised Theses or Diss.: Yes **Manuscript Copies:** 1
Review Period: 1-2 months (PR) **Lagtime:** 1-6 months
Types of Articles: Applications, research results, and project summaries.
Major Contents: Solar thermal, photovoltaics, high efficiency, and conservation.
Miscellaneous: This is a trade magazine, not a technical journal. Practical aspects should be emphasized in any submissions. Formerly: SOLAR ENGINEERING.

490 **SOLAR UTILIZATION NEWS (SUN)**

Alternate Energy Institute
P.O. Box 3100
Estes Park, CO 80517

Editor: Art Anderson
Phone: 303-586-5635

ISSN: 0736-0991 **Circulation:** 1,000 **Frequency:** M
Style Sheet: No
Revised Theses or Diss.: Yes **Manuscript Copies:** 1
Review Period: 2 weeks **Lagtime:** 2-3 months
Types of Articles: Research, projects, and book reviews.
Major Contents: Solar energy, earth sheltered housing, wind energy, conservation, and energy efficiency in housing.
Miscellaneous: Newsletter-type and format and information. Authors are rarely credited but contact people for projects are provided. Conversational tone prevails.

491 **SOLID STATE COMMUNICATIONS: AN INTERNATIONAL JOURNAL**

Pergamon Press
Maxwell House, Fairview Park
Elmsford, NY 10523

Editor: Elias Burstein
Phone: 914-592-7700
ISSN: 0038-1098 **Circulation:** 2,000 **Frequency:** 48/year
Style Sheet: Yes
Revised Theses or Diss.: Yes **Manuscript Copies:** 3
Review Period: 1 month (PR) **Lagtime:** 1-3 months
Types of Articles: Research, reviews, and communications.
Major Contents: Physical and chemical properties of solids and condensed systems.
Miscellaneous: A supplement to THE JOURNAL OF PHYSICS AND CHEMISTRY OF SOLIDS. As such it accepts short communications which require the maximum speed of publications.

492 **SOLID STATE IONICS**

Elsevier Science Publishers B.V.
North-Holland Physics Publishing Division
P.O. Box 211
1000 AE Amsterdam, The Netherlands

Editor: M. Stanley Whittingham, Rock Science, Schlumberger - Doll Research, P.O. Box 307, Ridgefield, CT 06877
ISSN: 0167-2738 **Circulation:** N/A **Frequency:** M
Style Sheet: No
Revised Theses or Diss.: N/A **Manuscript Copies:** 1
Review Period: N/A (PR) **Lagtime:** N/A
Types of Articles: Experimental and theoretical research, reviews, and relevant symposium proceedings.

Major Contents: The physics, chemistry and materials science of ion transport phenomena in solids, and properties resulting from ion mobility. The following constitute papers on topics such as: physics and chemistry of defects; ionic transport measurements and mechanisms; theories of ion transport; incorporation of ions into electronic conductors; many aspects of solid state electro-chemistry/physics including interfacial properties; processes such as corrosion, oxidation, sintering and other solid state reactions. Related technological applications such as batteries, fuel cells, electrochromic displays, solid state chemical sensors, hydrogen storage materials, catalysis, photography, electrolysis cells, metal winning, etc. will also be included.

493 **SOLID-STATE ELECTRONICS**
Pergamon Press
Maxwell House, Fairview Park
Elmsford, NY 10523

Editor: W. Crawford Dunlap
Phone: 914-592-7700
ISSN: 0038-1101 **Circulation:** 2,400 **Frequency:** M
Style Sheet: Yes
Revised Theses or Diss.: N/A **Manuscript Copies:** 2
Review Period: 1-2 months **Lagtime:** 8-10 months
Types of Articles: Research, reviews, notes, communications, and letters to the editors.
Major Contents: Topics relating to applied solid-state physics such as: transistor technology, transistor theory and design, crystal growing and handling, and transistor process technology.

494 **SOUTHEASTERN GEOLOGY**
Duke University, Geology Department
Box 6729, College Station
Durham, NC 27708

Editor: Dr. S. Duncan Heron, Jr.
Phone: 919-684-2206
ISSN: 0038-3678 **Circulation:** 700 **Frequency:** Q
Style Sheet: Yes
Revised Theses or Diss.: Yes **Manuscript Copies:** 2
Review Period: 2-3 months (PR) **Lagtime:** 6 months
Types of Articles: Research.
Major Contents: Geological research relating to the southeastern region of the U.S.

495 **SOUTHEASTERN OIL REVIEW**

Review Publishing Company
P.O. Box 145
Jackson, MS 39205

Editor: J. Ishee
Phone: 601-353-6213
ISSN: N/A **Circulation:** 4,800 **Frequency:** W
Style Sheet: No
Revised Theses or Diss.: No **Manuscript Copies:** 1
Review Period: 2 weeks **Lagtime:** 2 weeks
Types of Articles: Research, reviews, summaries, and news items.
Major Contents: New locations; drilling progress reports and well completions for oil and gas wells in southeastern states area, and drilling and exploration technologies.
Miscellaneous: Maps are accepted but pictures are not.

496 **SPACE SOLAR POWER REVIEW**

Pergamon Press
Maxwell House, Fairview Park
Elmsford, NY 10523

Editor: J.W. Freeman, Space Solar Power Review, Rice University, P.O. Box 1892, Houston, TX 77251
Phone: 713-527-8101
ISSN: 0101-9067 **Circulation:** 1,000 **Frequency:** Q
Style Sheet: No
Revised Theses or Diss.: Yes **Manuscript Copies:** 3
Review Period: 2 months (PR) **Lagtime:** 6 months
Types of Articles: Research, reviews, and news.
Major Contents: Space solar power or energy systems and related topics including environmental, social, institutional, engineering, microwave effects, and atmosphere or ionosphere effects affecting the feasibility and practicality of the solar power satellite concept.
Miscellaneous: Published for the Sunset Energy Council.

497 **STATISTICAL JOURNAL OF THE UNITED NATIONS COMMISSION FOR EUROPE**

Elsevier Science Publishers B.V.
P.O. Box 1991
1000 BZ Amsterdam
The Netherlands

Editor: Andreas Kahnert, Editor in Charge, ECE Statistical Division, Palais Les Nations, CH0-1211 Geneva 10, Switzerland
Phone: 022-31-02-11
ISSN: 0167-8000 **Circulation:** N/A **Frequency:** Q
Style Sheet: Yes
Revised Theses or Diss.: No **Manuscript Copies:** 4
Review Period: Varies (PR) **Lagtime:** 4 months
Types of Articles: Research, projects, reviews, and articles that have a recognizable link to the work program of the conference of European statisticians.
Major Contents: Regional statistical co-operation, organization and operation of statistical services, methods in economic statistics, social and demographic statistics, environmental statistics, electronic data processing, and energy statistics.

498 **STEAM AND FUEL USERS JOURNAL (INDIA)**
Steam and Fuel User's Association of India
4 B 1, J.P. Tower, 1, Dr. Thirumoorthy Road
Nungambakkam, Madras 600 034
India

Editor: J. Mathew
Phone: 478004 Madras
ISSN: 0039-0828 **Circulation:** 1,382 **Frequency:** Q
Style Sheet: No
Revised Theses or Diss.: Yes **Manuscript Copies:** 2
Review Period: 1 month **Lagtime:** N/A
Types of Articles: Research, projects, review case studies, technical articles, and developments.
Major Contents: Energy, steam, fuel, energy conservation, electricity, refrigeration, air pollution, water treatment, industrial process problems, and energy economics.

499 **SUGAR Y AZUCAR**
Ruspam Communications, Inc.
2050 Center Avenue
Ft. Lee, NJ 07024

Editor: Donald K. Luke
Phone: 201-461-8660
ISSN: 0039-4742 **Circulation:** 7,920 **Frequency:** M
Style Sheet: No
Revised Theses or Diss.: Yes **Manuscript Copies:** 2

Review Period: 2 months **Lagtime:** 2-6 months
Types of Articles: Research, projects, reviews, and actual case studies preferred.
Major Contents: All areas of the sugar industry, field and factory, including by-products.
Miscellaneous: Articles are in English and/or Spanish.

500 **SUN TIMES**
Solar Lobby, Suite 510
1001 Connecticut Avenue, N.W.
Washington, DC 20036

Editor: Joan Moody
Phone: 202-466-6350
ISSN: 0278-582X **Circulation:** 40,000 **Frequency:** BM
Style Sheet: N/A
Revised Theses or Diss.: N/A **Manuscript Copies:** N/A
Review Period: N/A **Lagtime:** N/A
Types of Articles: Brief informational pieces, and book reviews.
Major Contents: Renewable energy and conservation, politics, technology, legislation, community action, consumer information.
Miscellaneous: Does not accept unsolicited articles for Sun Times but does publish occasional unsolicited papers. These are normally 32 pages in length and deal with an unusual or exciting area of renewable energy or conservation. Contact the editor regarding such a publishing opportunity. Formerly: SUN DAY TIMES.

501 **SUN WORLD**
Pergamon Press
Maxwell House, Fairview Park
Elmsford, NY 10523

Editor: Dale A. Fousel, Sun World, P.O. Box 886, Berkeley, CA 94701
ISSN: 0149-1938 **Circulation:** 11,000 **Frequency:** Q
Style Sheet: No
Revised Theses or Diss.: N/A **Manuscript Copies:** N/A
Review Period: 2-4 months **Lagtime:** N/A
Types of Articles: Original research, articles, editorials, news comments, and book reviews.
Major Contents: Information and ideas relating to the worldwide development and use of solar energy.
Miscellaneous: Journal of the International Solar Energy Society, P.O. Box 26, Highett, Victoria 3 190, Australia. Send manuscripts directly to editor (above).

502 **SURFACE TECHNOLOGY**

Elsevier Sequoia S.A.
P.O. Box 851
CH-1001 Lausanne 1
Switzerland

Editor: J.P.G. Farr
ISSN: N/A **Circulation:** N/A **Frequency:** M
Style Sheet: See Miscellaneous
Revised Theses or Diss.: N/A **Manuscript Copies:** 3
Review Period: 1 month (PR) **Lagtime:** 9 months
Types of Articles: Original papers, invited reviews, brief contributions, book reviews, and notices of meetings.
Major Contents: Surface technologies such as: characterization of surfaces, surface preparation, electrodeposition of metals and alloys, the structure of electro-deposits, anodic films, vapor-deposited layers, and epitaxy.
Miscellaneous: Style guide may be obtained from the publishers.

503 **TECHNOLOGY REVIEW**

Massachusetts Institute of Technology
Room 10-140
Cambridge, MA 02139

Editor: John I. Mattill, Editor-in-chief
Phone: 617-253-8250
ISSN: 0040-1692 **Circulation:** 75,000 **Frequency:** M
Style Sheet: No
Revised Theses or Diss.: Yes **Manuscript Copies:** 1
Review Period: 1-2 months **Lagtime:** 2-3 months
Types of Articles: Research, projects, and review articles on new developments with emphasis on policy issues and implications.
Major Contents: All fields of energy technology and policy.

504 **TECTONOPHYSICS**

Elsevier Science Publishers B.V.
P.O. Box 211
1000 AE Amsterdam
The Netherlands

Editor: M. Friedman, Texas A & M University, College Station, TX 77843
Phone: 212-916-1250

ISSN: 0040-1951 **Circulation:** N/A **Frequency:** SA
Style Sheet: Yes
Revised Theses or Diss.: N/A **Manuscript Copies:** 3
Review Period: Varies **Lagtime:** Varies
Types of Articles: Research, reviews, and short papers.
Major Contents: Original studies and comprehensive reviews in the field of geotectonics and the geology and physics of the earth's crust and interior.

505 THIN SOLID FILMS
Elsevier Science Publishers
52 Vanderbilt Avenue
New York, NY 10017

Editor: J.N. Zemel, Center for Chemical Electronics,, Moore School of Electrical Engineering D-2, University of Pennsylvania, Philadelphia, PA 19104
Phone: 212-867-9040
ISSN: 0040-6090 **Circulation:** N/A **Frequency:** BM
Style Sheet: N/A
Revised Theses or Diss.: N/A **Manuscript Copies:** 3
Review Period: N/A **Lagtime:** N/A
Types of Articles: Research, reviews, letters, book reviews, announcements, conference reports, and news.
Major Contents: Electronics and optics, metallurgical and protective coatings, preparation and characterization, and general film behavior.
Miscellaneous: Manuscripts should be sent to Editorial Office (U.K.) of Elsevier Sequoia, Mayfield House, 256 Banbury Road, Oxford OX2 7DH, Great Britain.

506 TRANSPORTATION RESEARCH PART A: GENERAL
Pergamon Press
Maxwell House, Fairview Park
Elmsford, NY 10523

Editor: Frank A. Haight, The Pennsylvania Transportation Institute, Research Building B, The Pennsylvania State University, University Park, PA 16802
Phone: 914-592-7700
ISSN: 0191-2607 **Circulation:** 1,600 **Frequency:** BM
Style Sheet: Harvard System of References.
Revised Theses or Diss.: Yes **Manuscript Copies:** 3
Review Period: 2 months (PR) **Lagtime:** 6 months
Types of Articles: Research, case studies, survey, expository, and soft-

ware surveys.

Major Contents: Topics of general interest in all passenger and freight transportation modes: policy analysis, formulation, and evaluation; planning; interaction with the political, socioeconomic and physical environment; design, management and evaluation of transportation systems.

507 **TRANSPORTATION RESEARCH PART B: METHODOLOGICAL**

Pergamon Press
Maxwell House, Fairview Park
Elmsford, NY 10523

Editor: Frank A. Haight, The Pennsylvania Transportation Institute, Research Building B, The Pennsylvania State University, University Park, PA 16802
Phone: 914-592-7700
ISSN: 0191-2615 **Circulation:** 1,600 **Frequency:** BM
Style Sheet: Harvard System of References
Revised Theses or Diss.: Yes **Manuscript Copies:** 2
Review Period: 2 months (PR) **Lagtime:** 6 months
Types of Articles: Research, case studies, survey, expository, and software surveys.
Major Contents: Papers on all methodological aspects of transportation, especially those requiring mathematical techniques.

508 **TRANSPORTATION SCIENCE JOURNAL**

Operations Research Society of America
428 E. Preston Street
Baltimore, MD 21202

Editor: Denos C. Gazis
Phone: 301-528-4146
ISSN: 0049-4518 **Circulation:** 1,650 **Frequency:** Q
Style Sheet: Yes
Revised Theses or Diss.: N/A **Manuscript Copies:** 2
Review Period: 2 months **Lagtime:** 6 months
Types of Articles: Research, letters to the editor, book reviews, and announcements.
Major Contents: Original scientific contributions to the field of transportation science.

509 TRANSPORTATION THEORY AND STATISTICAL PHYSICS

Marcel Dekker, Inc.
270 Madison Avenue
New York, NY 10016

ISSN: N/A **Circulation:** N/A **Frequency:** N/A
Style Sheet: Yes
Revised Theses or Diss.: N/A **Manuscript Copies:** N/A
Review Period: 1 month (PR) **Lagtime:** 6-12 months
Types of Articles: Research.
Major Contents: Topics in irreversible statistical mechanics, or transport and kinetic theory.

510 TRANSPORTATION: AN INTERNATIONAL JOURNAL

Elsevier Science Publishers B.V.
P.O. Box 330
1000 AH Amsterdam
The Netherlands

Editor: Martin G. Richards
Phone: 212-867-9040
ISSN: N/A **Circulation:** N/A **Frequency:** Q
Style Sheet: Yes
Revised Theses or Diss.: N/A **Manuscript Copies:** N/A
Review Period: 1-2 months **Lagtime:** 6-8 months
Types of Articles: Research.
Major Contents: Transportation analysis, design and planning.

511 TULANE STUDIES IN GEOLOGY AND PALEONTOLOGY

Tulane University
Department of Geology
New Orleans, LA 70118

Editor: Hubert C. Skinner
Phone: 504-865-5198
ISSN: 0041-4018 **Circulation:** 1,000 **Frequency:** Q
Style Sheet: Yes
Revised Theses or Diss.: Yes **Manuscript Copies:** 2
Review Period: 3 months **Lagtime:** 12 months
Types of Articles: Research.
Major Contents: Geology and paleontology of the Gulf, Caribbean or general interest.

512 TURBOMACHINERY INTERNATIONAL MAGAZINE

Business Journals, Inc.
P.O. Box 5550
Norwalk, CT 06856

Editor: David N. Schaffer
Phone: 203-853-6015
ISSN: 0149-4147 **Circulation:** 12,000 **Frequency:** 9/year
Style Sheet: No
Revised Theses or Diss.: No **Manuscript Copies:** 1
Review Period: 2-4 weeks **Lagtime:** 2-4 months
Types of Articles: Case studies, maintenance practices, and installations.
Major Contents: Gas turbines; steam turbines; centrifugal, axial and screw compressors; turbo expanders; turbo pumps; cogeneration; combined cycles; hydro-turbines; wind turbines; accessory equipment (gears, couplings, filters, fuel pumps, etc.); components (blades, vanes, casings, nozzles, etc.); and power recovery.
Miscellaneous: Addressed to energy conversion managers and engineers. Formerly: GAS TURBINE INTERNATIONAL.

513 URJA

G-82 Sujan Singh Park
P.O. Box 3008
New Delhi 110003
India

Editor: Dipak B.R. Chaudhuri
ISSN: 0378-8535 **Circulation:** 4,600 **Frequency:** M
Style Sheet: Yes
Revised Theses or Diss.: Yes **Manuscript Copies:** 2
Review Period: 1 month **Lagtime:** 3-4 months
Types of Articles: Research, reviews, and comments.
Major Contents: Energy policy, planning and management; applied energy; petroleum exploration and exploitation; oil and gas; coal utilization; thermal power issues; energy environment interface; use of renewable resources and validity for developing countries.
Miscellaneous: International R&D which have applications to India and Asia are prevalent.

514 VERTICA: THE INTERNATIONAL JOURNAL OF ROTORCRAFT AND POWERED LIFT AIRCRAFT

Pergamon Press, Maxwell House
Fairview Park
Elmsford, NY 10523

Editor: P. Friedman
Phone: 914-592-7700
ISSN: 0360-5450 **Circulation:** 1,050 **Frequency:** Q
Style Sheet: Yes
Revised Theses or Diss.: N/A **Manuscript Copies:** 2
Review Period: 2 months (PR) **Lagtime:** 6-8 months
Types of Articles: Research, technical notes, reviews, letters to the editors, book reviews, and calendar of events.
Major Contents: Topics in the design and use of rotorcraft and aircraft that use engine thrust for lift production.

515 **WARTA GEOLOGI**
Geological Society of Malaysia
University of Malaya
Department of Geology
Kuala Lumpur 22-11, Malaysia

Editor: Dr. Teh Guan Hoe
Phone: Malaysia 03-577036
ISSN: 0126-5539 **Circulation:** 20,000 **Frequency:** BM
Style Sheet: Yes
Revised Theses or Diss.: Yes **Manuscript Copies:** 2
Review Period: 2 months **Lagtime:** 2 months
Types of Articles: Research and projects.
Major Contents: General topics related to geology, soils, and petroleum of southeast Asia.

516 **WATER RESEARCH**
Pergamon Press
Maxwell House, Fairview Park
Elmsford, NY 10523

Editor: Dr. F.G. Poland (USA Editor), School of Civil Engineering, Georgia Institute of Technology, Atlanta, GA 30332
Phone: 914-592-8727
ISSN: 0043-1354 **Circulation:** 3,800 **Frequency:** M
Style Sheet: Yes
Revised Theses or Diss.: Yes **Manuscript Copies:** 2
Review Period: N/A (PR) **Lagtime:** 6 months
Types of Articles: Research and reviews.
Major Contents: Water technology, chemistry, microbiology, the operation of processes for the treatment of municipal sewage and industrial waste waters, economic studies on pollution control methods, urban

storm drainage, studies on the treatment of sludges, the fate of bacteria and viruses in sewage disposal into the sea; water pollution problems in developing countries, and pollution of bathing beaches.

Miscellaneous: An official journal of the International Association on Water Pollution Research and Control. Manuscripts should be mailed to IAWPR Alliance House, 29-3 High Holborn, London WC1V 6BA, England.

517 **WAVE MOTION**

Elsevier Science Publishers B.V.
P.O. Box 211
1000 AE Amsterdam
The Netherlands

Editor: J.D. Achenbach, Editor-in-Chief, The Technological Institute, Department of Civil Engineering, Northwestern University, Evanston, IL 60201

ISSN: 0165-2125 **Circulation:** N/A **Frequency:** Q

Style Sheet: Yes

Revised Theses or Diss.: N/A **Manuscript Copies:** 2

Review Period: 2 months (PR) **Lagtime:** 6 months

Types of Articles: Research, reviews, letters, and discussions of previously published papers.

Major Contents: Topics in wave propagation phenomena in acoustics, optics, electromagnetic theory, and solid or fluid mechanics.

Miscellaneous: The pamphlet, "Instruction to Authors", is available on request from the publishers.

518 **WIND ENGINEERING**

Multi-Science Publishing Company, Ltd.
42/45 New Broad Street
London EC2M 1QY
England

Editor: Dr. E. Mowforth, Department of Mechanical Engineering, University of Surrey, Guildford, Surrey GU2 5XH, United Kingdom

Phone: London 01-628-0898

ISSN: 0309-524X **Circulation:** N/A **Frequency:** Q

Style Sheet: Yes

Revised Theses or Diss.: Yes **Manuscript Copies:** 2

Review Period: 1 month (PR) **Lagtime:** 3 months

Types of Articles: Research, projects, and reviews.

Major Contents: All aspects of wind energy systems: including wind-assisted marine propulsion, measurement methods, and related eco-

nomic and environmental topics.
Miscellaneous: An official journal of the European Wind Energy Association.

519 **WOOD'N ENERGY**
New Hampshire Times
P.O. Box 2008
Laconia, NH 03247

Editor: Kenneth Daggett
Phone: 603-528-4285
ISSN: 0273-5695 **Circulation:** 52,000 **Frequency:** M
Style Sheet: Yes
Revised Theses or Diss.: Yes **Manuscript Copies:** 1
Review Period: 1 month (PR) **Lagtime:** 2 months
Types of Articles: Research, projects, and reviews.
Major Contents: Wood and coal for residential heating, solid fuels, solar appliances, technology, and market studies and equipment.

520 **WORLD OIL**
Gulf Publishing Company
3301 Allen Parkway
Houston, TX 77001

Editor: T.R. Wright, Jr., WORLD OIL, P.O. Box 2608, Houston, TX 77001
Phone: 713-529-4301
ISSN: 0043-8790 **Circulation:** 40,000 **Frequency:** M
Style Sheet: Yes
Revised Theses or Diss.: Yes **Manuscript Copies:** 1
Review Period: 1 month (PR) **Lagtime:** 2-4 months
Types of Articles: Prefer practical operational type articles on exploration, drilling and production. Would not accept articles describing complex research methods. Articles reviewing available technology and its application will be considered.
Major Contents: Exploration, drilling, production, petroleum economics, oil and gas supply/demand forecasts; and oil business activity forecasts.
Miscellaneous: Prefer to work in English units rather than metric. If paper uses metric, English units must be included.

521 **WORLD PETROLEUM LEGISLATION**
Barrows Company, Inc.
P.O. Box 1591, Grand Central Station
New York, NY 10163

Editor: G.H. Barrows, Barrows 1B, 116 E. 66th Street, New York, NY 10021
Phone: 212-772-1199
ISSN: N/A **Circulation:** 500,000 **Frequency:** W
Style Sheet: Chicago Manual of Style.
Revised Theses or Diss.: Yes **Manuscript Copies:** 1
Review Period: 3-6 months (PR) **Lagtime:** 3-6 months
Types of Articles: Research and reviews.
Major Contents: Articles on petroleum legislation in the international field. Taxation of oil and gas, legislation lation relevant ot oil and gas, proposed changes in oil and gas legislation, reviews of articles and book in the field of oil and gas legislation around the world. Specialized articles on subjects such as offshore legislative requirements, types of organization dealing with computerization of legal materials in the energy field, legal and tax reviews.

522 ZEITSCHRIFT FUR GEOMORPHOLOGIE ANNALS OF GEOMORPHOLOGY/ANNALES

Gebrueder Borntraeger
Verlagsbuchand
Johan 3A 7000
Stuttgart 1, West Germany

Editor: Frau Professor Dr. Hanna Bremer, Geographology, Institute der University Koln, Albertus-Magnus-Platz, 5000 Koln 4
Phone: 0221-4702547
ISSN: N/A **Circulation:** N/A **Frequency:** Q
Style Sheet: Yes
Revised Theses or Diss.: No **Manuscript Copies:** 2
Review Period: 2-4 months (PR) **Lagtime:** 4-8 months
Types of Articles: Research.
Major Contents: Geomorphology.

APPENDIX I

JOURNALS SURVEYED BUT EXCLUDED

The following list is included to provide the researcher with an idea of the coverage of journals and organizations surveyed but which did not meet the requirements for inclusion. The "Scope and Selection Methodology" section of the PREFACE discusses deletions in depth.

ACADEMIE POLONAISE DES SCIENCES
Bulletin Serie Des Science De La Terre
Panstwowe Wydawnictwo Naukowe,
Miodowa 10, Warsaw,
Poland

ACTA GEOLOGICA HISPANICA
Instituto de Investigaciones
Geologicas Jaime Almera
Egipciacas 15
Barcelona-1, Spain

ACTA GEOLOGICA POLONICA
Panstwowe Wydawnictwo Naukowe
U1 Miodowa 10, Warsaw, Poland

ACTA GEOLOGICAL
Akademiai Kiado, Publishing House
P.O. Box 24
H-1363 Budapest
Hungary

ACTION OIL AND ENERGY REPORT
Action Oil and Energy Report, Inc.
219 Industrial Parkway
Lafayette, LA 70508

ADVANCED SOLAR ENERGY TECHNOLOGY NEWSLETTER
1609 West Windrose
Phoenix, AZ 85029

ALASKA MINES AND GEOLOGY
Division of Geological and
Geophysical Survey
P.O. Box 80007
College, AK 99708

AMERICAN NUCLEAR SOCIETY TRANSACTIONS
American Nuclear Society
555 N. Kensington Avenue
LaGrange Park, IL 60525

AMERICAN PETROLEUM INSTITUTE
Division of Statistics and Bulletin
2101 L Street, N.W.
Washington, DC 20037

APPLIED SOLAR ENERGY
Allerton Press, Inc.
150 Fifth Avenue
New York, NY 10011

ARAB OIL
Printing, Publishing and Distribution House
Box 2270
Airport Road
Shuwaikh, Kuwait

ARAB OIL REVIEW
No Forwarding Address

ARAB PETROLEUM
Arab Federation of Petroleum Mining and Chemical Workers
5 Zaki Street
Cairo, Egypt

ARCHIWUM MINERALOGICZNE
Polska Adademia Nauk
Wydawnictwa Geologiczne U1
Rakowiecka 4, Warsaw
Poland

ASHLAND NEWS
Ashland Oil, Inc.
P.O. Box 391
Ashland, KY 41114

ASOCIACION GEOLOGICA ARGENTINA REVISTA
Librart s.r.1
Casilla Correo 5047
Buenos Aires
Argentina

AUSTRALIA BUREAU OF MINERAL
Resouces, Geology and Geophysics
P.O. Box 378
Canberra City, ACT 2601, Australia

AUSTRALIAN ELECTRICAL WORLD
Strand Publishing Pty., Ltd.
432 Queen Street
Brisbane, Q1D 4000
Australia

AUSTRALIAN ELECTRONICS ENGINEERING
Thomson Publications Pty., Ltd.
47 Chippen Street
Chippendale, N.S.W.
2008, Australia

AWARE
Forwarding order has expired

BIOMASS DIGEST
A Publication of Technical Insights
P.O. Box 1304
Fort Lee NJ 07024

BMR JOUR. OF AUSTRALIAN GEO/ GEOPHY.
Bureau of Mineral Resources
Box 378
Canberra City
A.C.T., 2601, Australia

BREVOIRA GEOLOGICA ASTRUICA
Universidad de Oviedo
Instituto de Geologia Aplicada
Oviedo, Spain

BUREAU OF MINES
U.S. Department of the Interior
2401 E. Street, N.W.
Washington, DC 20241

BUSINESS AND SOCIETY REVIEW
Warren Gorham, and Lamont, Inc.
870 7th Avenue
New York, NY 10019

BUTANE PROPANE
197 F. Societe J A M
7 Square de Chatillon
75014 Paris, France

C G S NEWS
Bitech Publishers, Ltd.
1281 W. Georgia Street
Suite 100
Vancouver B.C. V6E 3J7, Canada

CAMERON SYMTHETIC FUELS REPORT
Rocky Mountain Div., Suite 400
650 South Cherry Street
Denver CO 80222

CANADIAN CHEMICAL PROCESSING
Southam Communications, Ltd.
1450 Don Mills Road
Don Mills, Ontario
M3B 2X7, Canada

CANADIAN ENERGY NEWS
6 Garamond Court
Don Mills, Ont. M3C 1Z5, Canada

CERT REPORT, THE
1000 Connecticut Avenue, N.W.
Suite 610
Washington, DC 20036

CHEMICAL MARKETING REPORTER
Schnell Publishing Company, Inc.
100 Church Street
New York, NY 10007

CHEMISTRY AND TECHNOLOGY OF FUELS AND OILS
Consultants Bureau
233 Spring Street
New York, NY 10013

COAL AGE
McGraw-Hill Publishing Company
1221 Avenue of the Americas
New York, NY 10020

COAL AND ENERGY QUARTERLY
Public Relations Department
Hobart House, Grosvenor Place
London
SW1 X7AE, England

COAL ENERGY NEWS
P.O. Box 538
St. Clairsville, OH 43950

CRC CRITICAL REVIEWS IN FOOD SCIENCE
CRC Press, Inc.
2000 Coyusocate Boulevard
Boca Raton, FL 33431

CRITICAL REVIEWS IN ENVIRONMENTAL CONTROL
2000 Northwest 24th Street
Boca Raton FL 33431

DEVELOPMENTS IN MINERAL PROCESSING
Elsevier Science Publishers B.V.
P.O. Box 211
1000 AE Amsterdam
The Netherlands

DRILL BIT
Hart Publications, Inc.
P.O. Box 1917
Denver, CO 80201

DRILLSITE
Maclean-Hunter Publishing Corporation
200-1015 Centre Street, N.
Calagary, Alta 2P8
Canada

E R I PERSPECTIVE
Energy Res. Inst. Information Ser.
University of Cape Town
Private Bag
Rondebosch 7700, South Africa

ECOLIBRIUM
Shell Oil Company
One Shell Plaza, P.O. Box 2463
Houston TX 77001

ECOLOGY OF FOOD AND NUTRITION
Department of Family Medicine, M.U.S.C.
171 Ashley Avenue
Charleston, SC 29425

ECONOMIE ELECTRIQUE/ ELECTRICAL SUPPLY
International Union of Producers and Distributors
39 Avenue de Friedland
75008 Paris, France

ECONOMIST, THE
The Economist Newspaper, Ltd.
P.O. Box 2700
Woburn, MA 01888

EGYPTIAN JOURNAL OF GEOLOGY
Geology Society of Egypt
24 Sharia Sherif Pasha
Research Department
Dokki, Cairo, Egypt

ELECTRA
International Conference on Large High Voltage
112 Boulevard
Haussman
75008 Paris, France

ELECTRIC TECHNOLOGY (USSR)
Pergamon Press
Maxwell House, Fairview Park
Elmsford, NY 10523

ELECTRICAL ENGINEER
Thomson Publications Pty., Ltd.
47 Chippen Street
Chippendale N.S.W.
2008, Australia

ELECTRICAL ENGINEER
Thomson Publications Pty., Ltd.
Box 8308
Johannesburg 2000
South Africa

ELECTRICAL INDIA
Chary Publications
14 Sidh Prasad, Ghatkopar Mahul Road
Tilak Nagar
Bombay 400089, India

ELECTRICAL POWER AND ENERGY SYSTEMS
Butterworth Scientific, Ltd.
Box 63, Westbury House
Bury Street
Guildford, Surrey GU2 5BH, England

ELECTRICAL POWER ENGINEER
Fox Lane North
Chertsey KT16 9HS
Surrey, England

ELECTRICITY AND ELECTRONICS
105 C New Alipore
Block F
Calcutta 700053
India

ELECTRICITY CANADA
Maclean-Hunter Publishers, Ltd.
Business Publishing Division
777 Bay Street
Toronto, Ontario M5W 1A7, Canada

ELECTRICITY CONSERVATION QUARTERLY
Devki Research and Development Engineers
Vasanrt Kutir
R.C. Dutt Road
Baroda 390 005, India

ELECTRO-OPTICS
Cahners Publishing Company
1350 E. Touhy Avenue, Box 5080
Des Plaines, IL 60018

ELECTRONIC ENGINEERING TIMES
M P Publications, Inc
111 E. Shore Road
Manhasset, NY 11030

ELECTRONIC MARKET TRENDS
Electronic Industries Association
Marketing Services Department
2001 Eye Street, N.W.
Washington, DC 20006

ELECTRONIC MODELING
Gordon and Breach Science Publishers, Ltd.
1 Bedford Street
London WC2E 9TP
England

ELECTRONIC PRODUCT DESIGN
Techpress Publishing Company, Ltd.
Walton House
93 High Street
Bromley BR1 1JW, England

ELECTRONICS
McGraw-Hill Publishing Company
1221 Avenue of the Americas
New York, NY 10020

ENERGETICOS
Direion General de Informacion
Secretaria de Patrimonio y Famento
Rio Rhin No. 22
Ler Piso Mexico 5 D.F., Mexico

ENERGIA E INDUSTRIA
Asociacion Venezolana de Ingenieria
Electrica y Mechanica
Apartado 6255
Caracas, Venezuela

ENERGY
Business Communications Company, Inc.
Box 2070-C, 9 Viaduct Road
Stamford, CT 06906

ENERGY
Objective Publications Pty. Ltd.
P.O. Box 584
Manuka
A.C.T. 2603, Australia

ENERGY CONSERVATION NEWS
Business Communications Company, Inc.
Box 2070-C, 9 Viaduct Road
Stamford, CT 06906

ENERGY DETENTE
P.O. Box 3996
North Hollywood, CA 91609

ENERGY DEVELOPMENTS
Un Bureau
Room 301
United Nations, NY 10017

ENERGY DIGEST(WATFORD, ENGLAND)
Comprint Ltd.
177 Hagden Lane
Watford, Herts
England

ENERGY MANAGEMENT
Penton-IPC
1111 Chester Avenue, Penton Plaza
Cleveland, OH 44114

ENERGY RESEARCH BUREAU
Fraser Management Associates
Box 494
Burlington, VT 05402

ENERGY TODAY
Trends Publishing, Inc.
National Press Building
Washington, DC 20045

ENGINEERING JOURNAL
American Institute of Steel Construction
400 North Michigan Avenue
Chicago, IL 60611

ENHANCED OIL-REC. FIELD REPORTS
Amer. Insti. of Min. Metal.& Engrs.
6200 N. Central Expressway
Dallas TX 75206

ETHOLOGY AND SOCIOBIOLOGY
Elsevier Science Publishers
52 Vanderbilt Avenue
New York, NY 10017

FIELDNOTES
Arizona Bureau of Geology and Mineral Technology
845 North Park Avenue
Tucson, AZ 85719

FOOD PROCESSING
Potman Publishing Company
430 N. Michigan Avenue
Chicago, IL 60611

FUEL ALCOHOL U.S.A.
Charles Walters Jr., Editor
Box 9547
Kansas City MI 64133

FUEL OIL AND OIL HEAT AND SOLAR SYSTEM
Industry Publications, Inc.
Commerce Road
Cedar Grove, NJ 07009

FUSION POWER REPORT
Business Publishers, Inc.
951 Pershing Drive
Silver Springs, MD 20910

GAS ENGINEERING AND MANAGEMENT
Institution of Gas Engineers
17 Grosvenor Crescent
London SW1X 7ES
England

GAS INDUSTRIES
Gas Industries Equipment and Appliance News, Inc.
P.O. Box 558
Park Ridge, IL 60068

GAS TURBINE WORLD
Pequot Publishing, Inc.
P.O. Box 494
Southport, CT 06490

GAS WORLD
Benn Publications
Sovereign Way, Tonbridge
Kent TN9 1RW, England

GASOHOL USA
10008 East 60th Terrace
Kansas City, MN 64133

GEOFISCA INTERNACIONAL
Universidad Nacional Autonoma
Instituto de Gofisica
Circuito Exter.
Mexico 20, D.F., Mexico

GEOLOGI
Geological Society of Finland
Kivimiehentie 1
SF-02150 Espools
Finland

GEOLOGICAL SOCIETY OF DENMARK BULLETIN
C.A. Reitzels Forlag
35 Noerre Soegade
DK-1370 Copenhagen K
Denmark

GEOLOGICAL SOCIETY OF JAPAN
Nihon Chishitsu Gakkai
Maruishi Bldg., 10-4 Kajicho
1-Chome
Chiyoda-ku, Tokyo 101 Japan

GEOLOGICAL SOCIETY OF THE PHILIPPINES JOURNAL
Bureau of Mines Building
Pedro Gil Street
Manila, Philippines

GEOLOGICAL SURVEY OF DENMARK
Denmarks Geologiske Undersoegelse
Norre Soegade 35, DK-1370
Copenhagen K, Denmark

GEOLOGICAL, MINING AND METALLURGICAL SOCIETY OF INDIA
Geology Department
University of Calcutta
35 B.C. Road
Calcutta 700019, India

GEOLOGIE EN MIJNBOUW/ GEOLOGY AND MINING
Royal Geological and Mining Society
Box 157
2000 AD Haarlem
The Netherlands

GEOLOGISKA FOERENINGENS I STOCKHOLM FOERHANDLINGER
Publishing House, Swedish Research
Box 6710
S-113 85 Stockholm
Sweden

GEOTHERMAL WORLD JOURNAL
International Society for Geothermal Engineers, Inc.
Drawer 4743
Whittier, CA 90607

GULF COAST OIL REPORTER
660 Bannock Street
Suite 100, Box 1917
Denver, CO 80201

HALF-YEARLY BULLETIN OF ELECTRICAL ENERGY STATISTICS FOR EUROPE
Palais des Nations
1211 Geneva 10
Switzerland

HEAT PIPE TECHNOLOGY
University of New Mexico
Technical Application Center
Albuquesque NM 87131

HOME ENERGY DIGEST
8009 34th Avenue, S.
Suite 170
Minneapolis, MN 55420

HOUSTON BUSINESS JOURNAL
Cordovan Corporation
5314 Bingle Road
Houston, TX 77092

HUDSON VALLEY GREEN TIMES
Grass Roots Energy and
Environmental Network
P.O. Box 208
Red Hook, NY 12571

I S G E TRANSACTIONS
International Society for Geothermal
Engineers, Inc.
Drawer 4743
Whittier, CA 90607

IEEE ELECTRON-DEVICE TRANSACTION
Institute of Electrical and Electronics
Engineers
345 E. 47th Street
New York, NY 10017

ILLUMINATING ENGINEERING INSTITUTE OF JAPAN JOURNAL
Shomei Gakkai
1-7-1 Yuraku-cho
Chiyoda-ku
Tokyo 100, Japan

IMPERIAL OIL REVIEW
Imperial Oil, Ltd.
111 St. Clair Avenue, W.
Toronto, Ontario
M5W 1K3, Canada

INDIAN GEOPHYSICAL UNION JOURNAL
Indian Geophysical Union
Hyderabad 7, India

INDUSTRIAL HEATING
National Industrial Pub. Com.
1610 Potomac Avenue
Pittsburgh PA 15216

INDUSTRIE DU PETROLE-GAZ-CHIMIE
Editions Olivier Lesourd
66 rue de la Rochefoucauld
75009 Paris, France

INFORM NEWS
Inform
381 Park Avenue, S.
New York, NY 10016

INSTITUTION OF ENERGY JOURNAL
Institute of Energy
18 Devoshire Street
Portland Place
London W. 1, England

INTERNATIONAL ASSOCIATION FOR MATHEMATICAL GEOLOGY
233 Spring Street
New York, NY 10013

INTERNATIONAL BUTANE/PROPANE NEWSLETTER
P.O. Box 419
Arcadia, CA 91006

INTERNATIONAL ENERGY STATISTICAL REVIEW
Central Intelligence Agency
Washington, DC 20505

INTERNATIONAL GEOLOGY REVIEW
Allen Press, Inc.
P.O. Box 368
Lawrence, KS 66044

INTERNATIONAL JOURNAL OF COMPUTER AND INFORMATION SCIENCES
Plenum Publishing Corporation
233 Spring Street
New York, NY 10013

INTERNATIONAL JOURNAL OF POWDER METALLURGY AND POWER TECHNOLOGY
The American Powder Metallurgy Institute
P.O. Box 2054
Princeton, NJ 08540

INTERNATIONAL JOURNAL OF RADIATION ENGINEERING
Jacob Slonim Publishing, Ltd.
P.O. Box 35
Tel-Aviv, Israel

INTERNATIONAL PRESS CUTTING SERVICE: PETROLEUM, PETROCHEMICAL,
FERTILIZATION, AGRICULTURE, AND CHEMISTRY
International Press Cutting Service
Box 63
Allahabad 211001, India

JOURNAL OF ARCHITECTURAL SCIENCE
University of Sidney
Department of Architectural Science
New South Wales 2006
Sidney, Australia

JOURNAL OF GEOMAGNETISM AND GEOELECTRICITY
Geophysical Institute
University of Tokyo, Faculty Science
2-11-16 Yayoi
Bunkyo-ku, Tokyo 113 Japan

JOURNAL OF NUCLEAR AGRICULTURE AND BIOLOGY
Indian Agriculture Research Institute
New Delhi 110012, India

JOURNAL OF PHYSICS OF THE EARTH
Center for Academic Publishers of Japan
2-4-16 Bunkyo-ku, Tokyo, 113 Japan

JOURNAL OF THE INSTITUTE OF ENERGY
Portland Place
18 Devonshire Street
London W1N 2AU
England

JOURNAL OF THE SCIENCE OF FOOD AND AGRICULTURE
Society of Chemical Industry
Balckwell Scientific Publications
Osney Mead, Oxfod
OX2 0EL, England

JOURNAL OF WIND ENERGY
American Solar Energy Society
110 West 34th Street
New York, NY 10001

KANSAS OIL MARKETER
Merchants National Bank
Suite 804
Topeka, KS 66612

KERNENERGIE
Akademie Verlag
Leipziger Street 3-4
1086 Berlin,
East Germany

LAND USE PLANNING REPORT
Business Publishers, Inc.
951 Pershing Drive
Silver Spring, MD 20910

LUNDBERG INWEEK
P.O. Box 3996
North Hollywood, CA 91609

LUNDBERG LETTER
P.O. Box 3996
North Hollywood, CA 91609

MATERIALS EVALUATION MAGAZINE
no forwarding address

MATERIALS EVALUATION MAGAZINE
Amer. Soc. for Nondestructive Test.
32000 Liverside Dr., Box 21142
Columbus OH 43221

MODERN POWER SYSTEMS
Eyre & Spottiswoode Publ., Ltd.
Swan House, 32 Swan Court
Leatherhead
Surrey KT22 8AH, England

MOUNTAIN GEOLOGIST
No forwarding address

NATIONAL PETROLEUM NEWS
Hunter Publishing Company
950 Lee Street
Des Plaines, IL 60016

NATURAL ENERGY AND LIVING
Natural Energy Association
35 Seplenham Road
Guildford
Surrey GY1 3RX, England

NATURAL GAS/NATURAL AND SYNTHETIC GAS
Forrestal Building
Washington, DC 20585

NATURE AND RESOURCES
Unesco
7-9 Place de Fontenoy
Paris 75700, France

NCRR BULLETIN: THE JOURNAL OF RESOURCE RECOVERY
1211 Connecticut Avenue, N.W.
Washington, DC 20036

NEW ROOTS
P.O. Box 340
Broomall PA 19008

NEW SCIENTIST
New Science Publications
813 National Press Building
Washington, DC 20045

NEWS FROM THE ENVIRONMENTAL COALITION ON NUCLEAR POWER
433 Orlando Avenue
State College, PA 16801

NLGI SPOKESMAN
National Lubricatring Grease Institute
4635 Wyandotte Street
Kansas City, MO 64112

NORTHEAST OIL REPORTER
Hart Publications, Inc.
660 Bannock Street, Suite 300
Denver, CO 80204

NUCLEAR CANADA/CANADA NUCLEAIRE
Canadian Nuclear Association
111 Elizabeth Street, 11th Floor
Toronto, Ontario
M5G 1P7, Canada

NUCLEAR FUEL CYCLE
U.S. Department of Energy
Box 62
Oak Ridge, TN 37830

NUCLEAR INDUSTRY
Atomic Industrial Forum, Inc.
7101 Wisconsin Avenue, N.W.
Washington, DC 20014

NUCLEAR INFO
Atomic Industrial Forum, Inc.
7101 Wisconsin Avenue, N.W.
Washington, DC 20007

NUCLEAR WASTE NEWS
Business Publishers, Inc.
951 Pershing Drive
Silver Spring, MD 20910

NUCLEUS
Pakistan Atomic Energy
Communication
P.O. Box 3112
Karachi-29, Pakistan
Karachi-29, Pakistan

OCEANS
Oceanic Society
Fort Mason
San Francisco, CA 94123

OFFSHORE ENGINEER
Thomas Telford, Ltd.
1-7 Great George Street
Westminister
London SWIP 3AA, England

OFFSHORE OIL WEEKLY
Morton Offshore Ltd., Surite D,
Silkhouse Court, Tithebarn St.
Liverpool
L2 2LZ, England

OFFSHORE SERVICES AND TECHNOLOGY
Kingston Publications, Ltd.
55/59 Fife Road
Kingston Upon Thames
Surrey KT1 1TA, England

OIL & ENERGY TRENDS
Energy Economics Research Ltd.
79 Queen Victoria St.
Reading Bucks
1SY England

OIL AND GAS REGULATION ANALYST
Executive Enterprising Publishing Company, Inc.
33 W. 60th Street
New York, NY 10023

OIL INDUSTRY COMPARATIVE APPRAISALS
John S. Herold Inc.
35 Mason Street
Greenwich, CT 06830

OIL PATCH: THE GULF COAST MAGAZINE
Oil Patch, Inc.
Box 53574
Lafayette, LA 70505

OIL REPORT
General Communications Inc.
899 Logan Street, Suite 600
Denver, CO 80203

OIL-LIFE STREAM OF PROGRESS
Caltex Petroleum Corporation
Box 61500
Dallas, TX 75261

PETROLEUM EQUIPMENT
Publex Corporation
P.O. 360
Whitehouse, NJ 08888

PETROLEUM OUTLOOK
John S Herold, Inc.
35 Mason Street
Greenwich, CT 06830

PETROLEUM SITUATION
Chase Manhattan Bank, Energy Economics Division
One Chase Manhattan Plaza
New York, NY 10005

POLLUTION ENGINEERING
no forwarding address

POLSKIE TOWARZYSTWO GEOLOGICZNE ROCZNIK SOCIETE GEOLOGIE DE
POLOGNE
Panstwowe Wydawnictwo Naukowe
Miodowa 10 Warsaw
Poland

PROGRESS
PIDC House
Dr. Ziauddin Ahmad Road
Karachi 4 Pakistan

PROGRESS IN PARTICLE AND NUCLEAR PHYSICS
Pergamon Press
Maxwell House, Fairview Park
Elmsford, NY 10523

QUARTERLY GEOLOGICAL NOTES
Department of Mines and Energy
191 Greenhill Road, Parkside
South Australia

RESEARCH MANAGEMENT
Technomic Publishing Company, Inc.
851 New Holland Avenue
New York, NY 10017

REVIEWS ON PETROLEUM CHEMISTRY
Freund Publishing House, Ltd.
P.O. Box 35010
Tel-Aviv, Israel

REVISTA ENERGETICA
Organ. Latinoamericana de Energia
Casilla 6313, Quito, Ecuador

REVUE DE L'INSTITUT FRANCAIS DU PETROLE
Service Publications
B.P. 311
92506 Rueil Malmaison
Cedex, France

ROCKS AND MINERALS IN CANADA
Consolidated Amethyst Communications, Inc.
60 Barbados Boulevard, Unit 6
Scarborough
Ontario M1J 1K9, Canada

S.U.N. - SOLAR UTILIZATION NEWS
Solar Utilization Network
121 Chestnut Street
Philadelphia, PA 19106

SCIENCE AND PUBLIC POLICY
IPC Business Press, Ltd.
Oakfield House, Perrymount Road
Haywards Heath
Sussex RH16, England

SCIENCE AND TECHNOLOGY FOR THE LEGISLATURE
Energy Program
1125 17th Street, Suite 1500
Denver, CO 80202

SCOTTISH JOURNAL OF GEOLOGY
Scottish Academic Press, Ltd.
33 Montgomery Street
Edinburgh EH7 5JX
Scotland

SCRIPTA FACULTATIS SCIENTIARUM NATURALIUM UNIVERSITATIS
PURKYNIANAE BRUNENSIS: BIOLOGIA
Prirodovedecka Fakulta
Kotlarska 2 611 37
Brno, *Cechoslovakia*

SCRIPTA FACULTATIS SCIENTIARUM NATURALIUM UNIVERSITATIS
PURKYNIANAE BRUNENSIS: CHEMIA
PRIRODOVEDECKA FAKULTA
KOTLARSKA 2 611 37
Brno, Czechoslovakia

SCRIPTA FACULTATIS SCIENTIARUM NATURALIUM UNIVERSITATIS
PURKYNIANAE BRUNENSIS: MATHEMATICA
Prirodovedecka Fakulta
Kotlarska 2 611 37
Brno, Czechoslovakia

SCRIPTA FACULTATIS SCIENTIARUM NATURALIUM UNIVERSITATIS
PURKYNIANAE BURNENSIS: PHYSICA
Prirodovedecka Fakulta
Kotlarska 2 611 37
Brno, Czechoslovakia

SHALE COUNTRY
Mountain Empire Publishing
200 Fillmore
Denver CO 80206

SIEMENS POWER ENGINEERING
Siemens AG, Redaktion
Postfach 3240, D-8520
Erlangen, W. Germany

SOCIETE BELGE DE GEOLOGIE BULLETIN
Belgische Verentiging Voor Geologie
13 rue Jenner
B-1040 Brussels
Belgium

SOCIETE GEOLOGIQUES DU NORD ANNALES
270 F. Societe Geologiques du Nord
BP 36, 59 Lille Gare, France

SOFT ENERGY NOTES
Friends of the Earth Foundation
1045 Sansome Street
San Francisco CA 94111

SOIL DYNAMICS AND EARTHQUAKE ENGR.
C M L Publications, Ashurst Lodge
Ashurst, Southampton SOE 2AA
England

SOLAR ENERGY DIGEST
74 W. Palo Verde Ave., Box 88
Ocotillo, CA 92259

SOLAR HEATING AND COOLING
Gordon Publications, Inc.
20 Community Place
Morristown NJ 07960

SOLAR LAW REPORTER
Conservation& Renewable Energy
P.O. Box 8900
Silver Springs MD 20907

STANFORD ENERGY REPORT
Stanford University, Terman
Engineering Center
Institute for Energy Studies
Stanford, CA 94305

STATE-OWNED ENERGY ENTERPRISES
UN Sectretariat Building
15 Washington Place, Room 301
New York, NY 10003

SUGAR JOURNAL
Sugar Journal, Inc.
4640 S. Carrollton Avenue, Suite 2-E
New Orleans, LA 70119

SUN-CLIPSE
University of Alabama in Huntsville
Environmental Energy Center
Huntsville, AL 35899

SUNSPOT
Univ. of Alabama in Huntsville
Kenneth E. Johnson Envir/Ener/Cent.
Huntsville AL 35899

SYNERJY
Box 4790, Grand Central Station
New York NY 10017

SYNFUELS WEEK
Pasha Publications
1401 Wilson Boulevard, Suite 910
Arlington, VA 22209

TEXAS ENERGY AND MINERAL RESOURCES
Center for Energy and Mineral
Resource
Texas A & M University
College Station TX 77843

TRANSACTIONS OF THE KANSAS ACADEMY OF SCIENCE
Kansas Academy of Science
2045 Constant Avenue, Campus West
Lawrence, KS 66044

U.S. GEOLOGICAL SURVEY
Professional Papers
12201 Sunrise Valley Drive
Reston, VA 22092

ULTRASONICS
IPC Sci. & Technology Press Ltd.
P.O. Box 63, Westbury House,
Bury Street
Guildford, Surrey Gu2 5BH, England

USTREDNI USTAV GEOLOGICKY VESTNIK
Publishing House of the
Czechoslovakia
Academy of Science
Vodickova 40
112 29 Prague 1, Czechoslovakia

WASHINGTON ATOMIC ENERGY REPORT
Reynolds Publishing Comapny, Inc.
Box 578
Glen Echo, MD 20768

WASTE AGE
Three Sons Publishing Company
6311 Gross Point Road
Niles IL 60648

WESTERN OIL REPORTER
Hart Publications
P.O. Box 1917
Denver, CO 80201

WIND INDUSTRY NEWS DIGEST
Alternative Sources of Energy (ASE)
107 S. Central Avenue
Milaca, MN 56353

WORLD COAL
Miller Freeman Publications
500 Howard Street
San Francisco, CA 94105

WORLD SOLAR MARKETS
Business Publishers, Inc.
951 Pershing Drive
Silver Spring, MD 20910

APPENDIX II

STYLE MANUALS FOR AUTHORS

Referenced throughout the "Master List" are style manuals. These as well as other selected style manuals and guides, are listed.

American Chemical Society. HANDBOOK FOR AUTHORS. Washington; 1978; ISBN 0-8412-0425.

American Mathematical Society. A MANUAL FOR AUTHORS OF MATHEMATICAL PAPERS. Providence, RI; 1984; ISBN 0-8218-0022-1.

American Society of Agronomy, Crop Science Society of America, and Soil Science Society of America. HANDBOOK AND STYLE MANUAL.Madison, WI: American Society of Agronomy; 1977; ISBN 0-89118-042-7.

American Geophysical Union. AGU STYLE GUIDE. Washington, D.C.; 1985.

American Institute of Physics. AIP STYLE MANUAL. New York; 1978.

American Society for Testing and Materials. ASTM STYLE MANUAL. Philadelphia; 1976.

American Society of Lubrication Engineers. ASLE AUTHOR'S GUIDE. Park Ridge, IL; 1977. Note: (SP-5).

American Society of Civil Engineers. AUTHOR'S GUIDE TO PUBLICATIONS OF ASCE. New York; 1983.

Bishop, Elna E. SUGGESTIONS TO AUTHORS OF THE REPORTS OF THE UNITED STATES GEOLOGICAL SURVEY. 6th Ed. Washington: U. S. Geological Survey; 1978.

Brightbill, George D. and Wayne C. Maxon. CITATION MANUAL FOR UNITED STATES GOVERNMENT PUBLICATIONS. Philadelphia, PA: Center for the Study of Federalism, Temple University; 1975.

Brusaw, Charles T. and Gerald J. Alred. HANDBOOK OF TECHNICAL WRITING. New York: St. Martin's Press; 1982; ISBN 0-312-35- 808-3.

Chandler, Harry E., (Ed.). TECHNICAL WRITERS HANDBOOK. Metals Park, OH: American Society for Metals; 1983; ISBN 0-87170- 151-0.

Cochran, Wendell et al. GEOWRITING: A GUIDE TO WRITING, EDITING & PRINTING IN EARTH SCIENCE. 3rd Ed. Falls Church, VA: American Geological Institute; 1979; ISBN 0-913312-13-4.

Council of Biology Editors Style Manual Committee. COUNCIL OF BIOLOGY EDITORS STYLE MANUAL: A GUIDE FOR AUTHORS, EDITORS, AND PUBLISHERS IN THE BIOLOGICAL SCIENCES. 5th Ed. Bethesda, MD; 1983.

DeBakey, Louis. THE SCIENTIFIC JOURNAL: EDITORIAL POLICIES AND PRACTICES. St. Louis, MO: Mosby; 1976; ISBN 0-8016-1223-3.

French, Christopher W. et al. ASSOCIATED PRESS STYLEBOOK AND LIBEL MANUAL: STYLE MANUAL FOR JOURNALISM. New York: The Associated Press; 1984; ISBN 0-201-10091-6.

GRAHAM & TROTMAN JOURNALS; INSTRUCTIONS TO CONTRIBUTORS. London: Graham & Trotman Limited; n.d.

A GUIDE FOR WILEY-INTERSCIENCE AND RONALD PRESS AUTHORS: IN THE PREPARATION AND PUBLICATION OF MANUSCRIPT AND ILLUSTRA TIONS. 2nd Ed. New York: Wiley; 1979.

Gulf Publishing Company. AUTHOR'S HANDBOOK: HYDROCARBON PROCESSING. Houston, TX; 1979.

Harkins, C. and D. L. Plung. A GUIDE TO WRITING BETTER TECHNICAL PAPERS. New York: Institute of Electrical and Electronics Engineers; 1981; ISBN 0-87942-157-6. Note: Pages 33-37 is an aid for selecting suitable publications for publishing.

Harris, John S. and Reed H. Blake. TECHNICAL WRITING FOR SOCIAL SCIENTISTS. Chicago, IL: Nelson-Hall; 1976.

HOW TO WRITE FOR CHEMICAL ENGINEERING. New York: McGraw-Hill; 1983.

Howell, John Bruce. STYLE MANUALS OF THE ENGLISH-SPEAKING WORLD: A GUIDE. Phoenix, AZ: The Oryx Press; 1983; ISBN 0-8977-089-0.

IEEE POWER ENGINEERING SOCIETY PUBLICATION GUIDE FOR POWER ENGINEERING. New York: Institute of Electrical and Electronics Engineers; 1981.

IFT STYLE GUIDE FOR RESEARCH PAPERS. Chicago, IL: Institute of Food Technologists; n.d.

INFORMATION FOR IEEE AUTHORS. New York: Institute of Electrical and Electronics Engineers; 1966.

Jordan, Lewis. NEW YORK TIMES MANUAL OF STYLE AND USAGE: STYLE MANUAL FOR THE FIELD OF JOURNALISM. New York: Times Book; 1977; ISBN 0-8129-0578-4.

Jordan, Stello et al, Ed. HANDBOOK OF TECHNICAL WRITING PRACTICES. New York: Wiley Interscience; 1971. Note: Wiley Series on Human Communication.

Leggett, Glenn, et al. PRENTICE-HALL HANDBOOK FOR WRITERS. 9th Ed. Englewood Cliffs: Prentice-Hall; 1975; ISBN 0-13-6950- 15-9.

Lonyear, Marie. THE MCGRAW-HILL STYLE MANUAL: CONCISE GUIDE FOR WRITERS AND EDITORS. New York: McGraw-Hill; 1982; ISBN 0-07-038676-5.

Mullins, Carolyn J. THE COMPLETE MANUSCRIPT PREPARATIONS STYLE GUIDE. Englewood Cliffs, NJ: Prentice Hall; 1982.

Mullins, Carolyn J. A GUIDE TO WRITING AND PUBLISHING IN THE SOCIAL AND BEHAVIORAL SCIENCES. Malabar, Fl: R.E. Krieger; 1983; ISBN 0-89874-643-4.

National Education Association. STYLE MANUAL FOR WRITERS AND EDITORS. New York: MacMillan; 1963.

Risley, Curtis. AYER PUBLIC RELATIONS AND PUBLICITY STYLE. Ft. Washington, PA: IMS Press; 1983 rev. Ed.; ISBN 0-901019- 0-27-5.

Rothman, Marie H. CITATION RULES AND FORMS FOR UNITED NATIONS DOCUMENTS AND PUBLICATIONS. Brooklyn, NY: Long Island University Press; 1971.

Schmidt, Steven. Creating the Technical Report. Englewood Cliffs, NJ: Prentice-Hall; 1983; ISBN 0-13-189027-1.

Sherman, Theodore A. and Simon S. Johnson. MODERN TECHNICAL WRITING. Englewood Cliffs, NJ: Prentice-Hall; 1983; ISBN 0-13-598748.

Skillin, Marjorie E. and R. Gay. WORDS INTO TYPE. Englewood Cliffs, NJ: Prentice-Hall; 1974; ISBN 0-13-964262-5.

Souther, James W. and Myron L. White. TECHNICAL REPORT WRITING. New York: Wiley Interscience; 1984; ISBN 0-89774-89-0.

Strunk, W., Jr., and E. B. White. THE ELEMENTS OF STYLE. 3rd Ed. New York: The Macmillan Company; 1979; ISBN 0-02-4181- 90-0.

A UNIFORM SYSTEM OF CITATION. Cambridge, MA: Harvard Law Review Association; 1976.

University of Chicago Press. CHICAGO MANUAL OF STYLE. Chicago; 1982; ISBN 0-226-10390-0.

Water Pollution Control Federation. AUTHOR'S GUIDE FOR JOURNAL WATER POLLUTION CONTROL FEDERATION. Washington; 1984. Note: Reprint from WPCF YEARBOOK.

Woodford, F. Peter (Ed.). SCIENTIFIC WRITING FOR GRADUATE STUDENTS. Bethesda, MD: Council of Biology Editors; 1982; ISBN 0-914340-01-8. Note: Prepared under the direction of the CBE Committee on Graduate Training in Scientific Writing.

APPENDIX III

ENERGY DATA BASE SUBJECT CATEGORIES AND SCOPE

The "Index" was developed using the U.S. Department of Energy's **ENERGY DATA BASE SUBJECT CATEGORIES AND SCOPE** as an authoritative guide to subject and index terms. The following is a copy of this DOE publication, reprinted with permission of the U.S. Department of Energy, Office of Scientific and Technical Information.

The subject scope of the Energy Data Base (EDB) encompasses all DOE-sponsored research. Broadly defined, EDB subject scope includes all technological aspects of energy production, conversion, and efficient utilization, and the economic, social, and political aspects as well. Scope notes are provided to define the extent of interest in certain subject areas, particularly areas of basic research. Cross references between categories are provided to aid both the categorization of information and its retrieval.

Citations entered into DOE's computerized bibliographic information system are assigned six-digit subject category numbers to broadly group information for storage, retrieval, and manipulation. These numbers are used in the preparation of printed documents, such as bibliographies and abstract journals, to arrange the citations and to aid searching on the DOE/RECON on-line system. This document has been prepared for use by (1) those individuals responsible for the assignment of category numbers to documents being entered into the Office of Scientific and Technical Information (OSTI) system, (2) those individuals and organizations processing magnetic tape copies of the files, (3) those individuals doing on-line searching for information in OSTI-created files, and (4) others who, having no access to RECON, need a printed copy.

The first column shows the six-digit numbers assigned to documents. The second column presents the category names and text to define the scope of interest.

Category Numbers

The six-digit category number is utilized as if it were three pairs of two-digit numbers. The first pair, as shown in the table of contents, designates a main or prime category. The middle pair designates a subset of the first pair, and the last or third pair designates a subset of the middle pair. Thus, in the category number 010406, the 01 designates COAL AND COAL PRODUCTS, the 04 designates **Processing** of coal and coal products, and the 06 designates

Office of Scientific and Technical Information
United States Department of Energy, P. O. Box 62, Oak Ridge, TN 37831

Pyrolysis and Cracking methods of processing coal and coal products. There are 39 first-level, 293 second-level, and 497 third-level categories; the category 06, FUSION FUELS, is no longer valid.

Category Use

Categories are used to subset the computer-stored files during search and retrieval operations. As in any other subsetting operation, this allows both greater precision and machine efficiency.

Categories are also used in generating the tables of contents and arranging citations in publications. In most publications the categories appear in numeric order, but corresponding sequence numbers that do not appear in DOE/TIC-4584 allow other arrangements.

Some users of EDB tapes perform SDI functions or are interested in selected subject areas only. The categories may be used in these operations.

Category Types and Assignments

The primary categories of EDB may be classified into three general types: (1) those representing devices, e.g., Nuclear Power Plants and Advanced Automotive Propulsion Systems; (2) those representing energy sources, e.g., Petroleum and Geothermal Energy; and (3) those representing scientific disciplines, e.g., Chemistry and Physics.

Any item processed into EDB is assigned at least one category; additional categories are assigned as needed to reflect the content and/or purpose of the document. In the case of multiple categories, the first category determines the location of the item in a publication and, therefore, should be selected most carefully; the additional categories are used as cross-references.

The user of this publication, as well as users of other OSTI publications or data bases, should remain aware of the two significant differences between categories and subject descriptors: (1) No effort is made to categorize to the level of specificity reached by subject descriptors. (2) Subject descriptors are used to reflect the document content, never or rarely the document purpose.

Index

An index is included to provide quick entry into the subject content of the category titles and category scope notes. The index is arranged alphabetically using the first word of a keyword phrase, with its corresponding category number and page reference.

ABOUT THE OFFICE OF SCIENTIFIC AND TECHNICAL INFORMATION

The Office of Scientific and Technical Information (OSTI) in Oak Ridge, Tennessee, has been the national center for scientific and technical information for the Department of Energy (DOE) and its predecessor agencies since 1946. In developing and managing DOE's technical information program, OSTI places under bibliographic control not only DOE-originated information but also worldwide literature on scientific and technical advances in the energy field and announces the source and availability of this information. Whereas the literature of science is emphasized, coverage is extended to DOE programmatic, socioeconomic, environmental, legislative/regulatory, energy analysis, and policy-related areas. To accomplish this mission, OSTI builds and maintains computerized energy-information data bases and disseminates this information via computerized retrieval systems and announcement publications such as abstracting journals, bibliographies, and update journals. Direct access to OSTI's most comprehensive data base, the Energy Data Base, is available to the public through commercial on-line bibliographic retrieval systems. The Energy Data Base and many of OSTI's energy-related data bases are available to DOE offices and contractors and to other government agencies via DOE/RECON, the Department's on-line information retrieval system. The Office of Scientific and Technical Information has developed and maintains systems to record and communicate energy-related research-in-progress information, to maintain a register of DOE public communications publications, to track research report deliverables from DOE contractors, and to test and make available DOE-funded computer software programs with scientific and management applications; OSTI also serves as adviser for special publication needs of the Department. To effectively manage DOE's technical information resources, OSTI's program is one of continual development and evaluation of new information products, systems, and technologies.

UNITED STATES DEPARTMENT OF ENERGY

John S. Herrington
Secretary

Martha Hesse Dolan
Assistant Secretary
Management and Administration

William S. Heffelfinger
Director of Administration

Joseph G. Coyne
Manager
Office of Scientific and Technical Information

David E. Bost
Editor

DOE/TIC-4584-R6
(DE84017376)
March 1985
Distribution Categories UC-2 and UC-13

CONTENTS
(Alphabetical Listing by Category Title)

CONTENTS

(Numerical Listing by Category Number)

CATEGORIES AND SCOPE NOTES

Category No.	Category Title
010000	COAL AND COAL PRODUCTS
010400	**Processing**
010401	*Carbonization* (Processes, operation of coke ovens and other coking equipment, and coke production statistics.)
010402	*Desulfurization and Purification* (Desulfurization, purification, and upgrading of coal, liquid and gaseous fuels made from coal, and other products made from coal. Information on the desulfurization, denitrification, or other cleaning of waste products such as flue gas from coal combustion, coal mining wastes, or waste emissions from coal processing plants is included under **Waste Management**.)
010403	*Hydrogenation*
010404	*Gasification* (Coal-gasification processes, kinetics, chemistry, products, economics, plant design, and constraints. Use other **Processing** categories when the total gasification process is not designated.)
010405	*Liquefaction* (Coal-liquefaction processes, kinetics, chemistry, products, economics, plant design, and constraints. Use other **Processing** categories when the total liquefaction process is not designated.)
010406	*Pyrolysis and Cracking*
010407	*Solvent Extraction*
010500	**By-Products** (Nonwaste by-products of the coal industry, such as organic chemicals from processing or use of coke in blast furnaces.)
010600	**Properties** (Physical and chemical properties of coal and coal products. Information on chemical analysis is also included.)
010800	**Waste Management** (Desulfurization or other purification of flue gas from coal combustion or burning of other coal-derived fuels, management of waste products from any coal-processing operation, and waste problems at coal mines.)
010900	**Environmental Aspects** (Environmental implications of developing, transporting, and using coal or coal products.)
011000	**Reserves and Exploration** (Reserves, exploration, and geology, including stratigraphy, mineralogy, and origin. For petrology, see the category **Properties**.)
012000	**Mining** (Mining equipment and techniques, mine planning and operation, and productivity.)
013000	**Transport and Handling** (All aspects of coal handling and transport, e.g., physical processes in coal preparation, coal pulverization, pipeline transport of coal and coal products, shipment statistics, etc.)
014000	**Combustion** (Efficient combustion of coal or coal products as well as coal combustion kinetics and mechanisms.)
015000	**Marketing and Economics** (Consumption, supply and demand, trade, charges, and factors affecting price.)
016000	**Health and Safety**
017000	**Legislation and Regulations** (Legislation, hearings, government policy statements, international agreements, etc.)

Category No.	Category Title
020000	**PETROLEUM** (General or survey papers that cannot be more specifically categorized, proposals for development, etc.)
020100	**Reserves** (Reserves and resources.)
020200	**Geology and Exploration** (Geophysical surveys, site geology, petroleum genesis; exploration techniques including well logging, geophysical techniques, and exploratory drilling.)
020300	**Drilling and Production** (All aspects of production drilling and techniques for increasing oil-field productivity, including secondary and tertiary methods. Exploratory drilling is categorized under **Geology and Exploration.**)
020400	**Processing** (All refining operations, including production of fuel oils.)
020500	**Products and By-Products** (Petrochemicals: only papers on products and by-products that are fuels or relate to fuel processing are included. Product recovery from wastes is included under **Waste Management.**)
020600	**Health and Safety**
020700	**Marketing and Economics** (Charges and factors affecting prices; supply and demand; trade; and consumption.)
020800	**Waste Management** (Desulfurization or other purification of flue gas from burning petroleum products; management of wastes during entire process from drilling through refining; and reclamation of waste lubricating oils and other petroleum products by re-refining.)
020900	**Environmental Aspects** (Effects of various steps in drilling, production, refining, and transporting petroleum and petroleum products. Oil spills are included. See also the **ENVIRONMENTAL SCIENCES** categories and **ADVANCED AUTOMOTIVE PROPULSION SYSTEMS, Emission Control.**)
021000	**Legislation and Regulation** (Legislation, hearings, government policy statements, international agreements, etc.)
022000	**Transport, Pipelines, and Handling**
023000	**Properties** (Chemical composition studies of petroleum from different fields; chemical analysis and composition studies of petroleum fuel products; chemical and physical properties of petroleum and petroleum fuel products.)
024000	**Storage** (Facilities, degradation of products during storage, and means of preventing degradation.)
025000	**Combustion** (Combustion of petroleum or petroleum products.)
030000	**NATURAL GAS** (General or survey papers that cannot be more specifically categorized, proposals for development, etc.)
030100	**Reserves** (Reserves and resources.)
030200	**Geology and Exploration** (Geophysical surveys, site geology, and natural gas genesis; exploration techniques including well logging, geophysical techniques, and exploratory drilling.)
030300	**Drilling, Production, and Processing** (All aspects of production, drilling, and processing for removal of impurities. Exploratory drilling is categorized under **Geology and Exploration.** Techniques for stimulating reservoir productivity are categorized under **Artificial Stimulation.**)
030400	**Products and By-Products** (Limited to papers on products and by-products that are fuels or relate to fuel processing, e.g., petrochemicals, LNG, petroleum gas liquids, etc. Product recovery from wastes is included under **Waste Management.**)
030500	**Health and Safety**
030600	**Marketing and Economics** (Charges and factors affecting price; supply and demand; trade; and consumption.)

Category No.	Category Title
030700	**Waste Management** (Management of wastes during entire process from drilling, production, and processing through end use.)
030800	**Environmental Aspects** (Effects of various steps in drilling, production, processing, and transporting, including LNG production, transport, and storage. LNG spills are included. See also the **ENVIRONMENTAL SCIENCES** categories and **ADVANCED AUTOMOTIVE PROPULSION SYSTEMS, Emission Control.**)
030900	**Artificial Stimulation** (Techniques for increasing gas-field productivity, including hydraulic fracture, underground nuclear explosions, and recovery from unconventional sources such as methane from coal seams or geopressured systems.)
031000	**Legislation and Regulation** (Legislation, hearings, government policy statements, international agreements, etc.)
032000	**Transport, Pipelines, and Handling**
033000	**Properties** (Studies on chemical composition of natural gas from different fields; chemical analysis; chemical and physical properties of natural gas and its fuel products.)
034000	**Combustion** (Combustion of natural gas or natural gas products.)
035000	**Storage** (Facilities, degradation of products during storage, and means of preventing degradation. Storage of LNG is included under this category.)
040000	OIL SHALES AND TAR SANDS (General or survey papers, proposals for development of industry, etc.)
040100	**Reserves and Exploration** (Reserves, resources, and exploration techniques, including geophysical surveys and exploratory drilling.)
040200	**Site Geology and Hydrology**
040201	*Site Geology* (Stratigraphy, lithology, geomorphology, petrology, mineralogy, and origin.)
040202	*Site Hydrology*
040300	**Drilling, Fracturing, and Mining** (Technology related to drilling, fracturing, and underground and surface mining.)
040400	**Oil Production, Recovery, and Refining**
040401	*In Situ Methods, True and Modified* (In-place recovery of oils and bitumens using combustion, chemical and steam stimulation, gasification, etc., and modified methods, which include limited surface processing.)
040402	*Surface Methods* (Methods used aboveground for recovery of oils and bitumens from mined sands and shales.)
040403	*Refining* (Upgrading oils from shales or sands.)
040404	*Purification* (Desulfurization, denitrification, etc.)
040500	**Properties and Composition** (Chemical composition, oil yield, chemical and physical properties of shales, sands, and their products.)
040600	**Direct Uses and By-Products** (Chemical and mineral extraction, use as boiler fuel, and productive use of spent shales or tar sand tailings.)
040700	**Health and Safety**
040800	**Marketing and Economics** (Factors affecting prices; supply and demand; trade; and consumption.)
040900	**Waste Research and Management** (Processing, control, or disposal of spent shales and other wastes.)
041000	**Environmental Aspects** (Studies on animal life, plant life, and cultural resources of affected area, surface- and groundwater pollution, site revegetation, and overall effects of disposal of spent shales, tar sand tailings, etc.)

Category No.	Category Title
042000	**Regulations**
050000	**NUCLEAR FUELS** (All information on nuclear fuels except design, assembly, and performance of fuel elements and assemblies, fuel handling procedures at reactors, fuel requirements, fuel burnup, and fuel-cycle economics. For this information, see **NUCLEAR POWER PLANTS** or **NUCLEAR REACTOR TECHNOLOGY**.)
050100	**Reserves** (Mineralogy, geology, properties, petrogenesis, resources, and estimated reserves of uranium and thorium minerals and ores.)
050200	**Exploration** (Identification and prospecting.)
050300	**Mining**
050400	**Feed Processing** (Chemical processing of ores for recovery of uranium or thorium, including all extraction, conversion, or reduction steps. Also includes design and chemical engineering of plants for such processing and process design, performance, and operation.)
050500	**Enrichment**
050501	*Gaseous Diffusion*
050502	*Centrifugation*
050503	*Laser Excitation*
050600	**By-Products**
050601	*Depleted Uranium*
050700	**Fuels Production and Properties** (Production, fabrication, and properties of fuels. See also specific **MATERIALS** and **CHEMISTRY** categories.)
050800	**Spent Fuels Reprocessing** (Reactor-fuel reprocessing, including analytical control, chemical separation, solvent studies, and plant and process design, performance, and operation.)
050900	**Transport and Storage** (Includes regulatory and administrative aspects. For design and development of shipping containers, see **ENGINEERING**. For storage of radioactive wastes, see *Waste Disposal and Storage*.)
051000	**Marketing and Economics**
052000	**Waste Management** (Studies, including economic, related to methods for the management, processing, disposal, or storage of radioactive wastes and nonradioactive wastes generated by nuclear facilities.)
052001	*Waste Processing* (Radioactive waste processing for concentration, decontamination, or fission-product recovery, includes plants, structures, and equipment. Regulatory and administrative aspects are also included.)
052002	*Waste Disposal and Storage* (All ultimate storage and disposal methods, e.g., land burial, sea disposal, tank storage, or salt-mine storage. Geological, hydrological, seismological, and nuclide-migration studies related to waste disposal are included. Monitoring, migration, and distribution of all radioactive materials in the atmosphere, soil, and waters of the Earth are included under the **ENVIRONMENTAL SCIENCES** categories.)
053000	**Environmental Aspects** (Environmental effects of various steps in the fuel handling and processing cycle, e.g., mining, enrichment, fabrication, or reprocessing. Includes radioactive-materials monitoring and migration when related to the fuel cycle. All monitoring, migration, and distribution of radioactive materials in the atmosphere, soil, and waters of the Earth are included under the **ENVIRONMENTAL SCIENCES** categories.)
054000	**Health and Safety** (Includes remedial action programs and activities.)
055000	**Regulations** (Legal aspects of all nuclear-safeguards issues, legal aspects of physical protection, and other legal aspects of the nuclear fuel cycle.)

Category No.	Category Title
055001	*Accountability and Safeguards* (Equipment, methods, standards, and systems for safeguarding and management of materials of nuclear interest, e.g., fissionable materials, radioactive materials, or materials of special interest such as heavy water. Research and development on equipment, methods, and systems to guard against the diversion of nuclear materials from uses permitted by law or treaty. Research and development of nuclear-materials accounting systems.)
055002	*Administrative and Regulatory* (Administrative, political, economic, organizational, and other aspects of the development and application of safeguards.)
060000	FUSION FUELS (*Note: no longer a valid category.* See **FUSION ENERGY,** subcategories 700205 and 700206.)
070000	**ISOTOPE AND RADIATION SOURCE TECHNOLOGY**
070100	**Physical Isotope Separation** (Both radioactive and stable isotopes. Includes design, performance, and efficiency of equipment for isotope separation, such as gas centrifuges, separation nozzles, etc. For uranium enrichment, see **NUCLEAR FUELS.** See also **CHEMISTRY,** *Radioisotope Production.*)
070101	*Heavy Water Production*
070200	**Radiation Sources** (Nuclear radiation sources only. Sources of other types of radiation, e.g., visible, monochromatic, microwave, etc., are categorized under more appropriate categories.)
070201	*Design and Fabrication*
070202	*Use in Food Processing*
070203	*Use in Medical Supply Sterilization*
070204	*Other Industrial Uses*
070300	**Isotopic Power Supplies** (Use of separated radioisotopes or mixed fission products as sources of electric, propulsive, or thermal energy. See also **DIRECT ENERGY CONVERSION, Thermoelectric Generators** and **Thermionic Converters.**)
070301	*Design and Fabrication*
070302	*Uses*
080000	**HYDROGEN** (General papers that cannot be more specifically categorized, e.g., papers on the hydrogen-based economy concept.)
080100	**Production** (General production reviews, separation processes, purification, and all processes for which there are no specific subcategories.)
080101	*Electrolysis* (Electrolytic decomposition of a chemical compound, usually water.)
080102	*Thermochemical Processes* (Water enters chemical reactions in which intermediate chemicals are recycled, and the overall process produces hydrogen and oxygen in separate steps.)
080103	*Steam Reformer Processes* (Volatile hydrocarbons are reacted with steam over a nickel catalyst at 700–1000°C to produce carbon oxides and hydrogen. The crude hydrogen is cooled and fed to a shift converter where it reacts with the carbon oxides and produces additional hydrogen.)
080104	*Water Gas Processes* (Water gas with steam in excess is passed over a catalyst, usually iron oxide, to produce hydrogen.)
080105	*Bosch Process* (Catalytic process for producing hydrogen from carbon monoxide and steam.)
080106	*Biosynthesis and Photochemical Processes* (Production of hydrogen by bacteria, algae, enzymes, photochemical methods, bioconversion, etc.)

Category No.	Category Title
080107	*Coal Gasification* (Production of hydrogen from coal, mainly by gasification.)
080108	*Steam-Iron Process* (Steam and air react with carbonaceous material to prepare a gas mixture containing carbon oxides and hydrogen which is used to reduce ferric oxide to ferrous oxide and iron. The iron is oxidized with steam to prepare hydrogen, and the cycle is repeated.)
080109	*Partial Oxidation Processes* (Partial oxidation of any liquid or gaseous hydrocarbon, including heavy fuel oils.)
080200	**Storage**
080201	*Chemisorption* (Storage as a hydride, primarily by sorption by metals.)
080202	*Underground* (Bulk storage in aquifers and geologic formations.)
080203	*Cryogenic* (Storage in tanks, dewars, pipelines, etc.)
080300	**Transport** (Distribution, materials handling, energy transmission, and transport through pipelines.)
080400	**Marketing and Economics** (Specific aspects of the economics of introducing hydrogen into the energy system.)
080500	**Safety** (All safety aspects of hydrogen production, storage, and use as fuel, including fires, explosions, etc.)
080600	**Industrial and Commercial Use** (Use of hydrogen as a fuel or as a raw material in industrial processes; also includes residential use of hydrogen and use of hydrogen as a transportation fuel.)
080700	**By-Products** (By-products, primarily of hydrogen production, and their use and economic impact.)
080800	**Properties** (Properties pertinent to the use of hydrogen as a fuel or energy carrier, such as effects on metals or other materials.)
080900	**Environmental Aspects** (Environmental implications of using hydrogen as an energy source.)
090000	**OTHER SYNTHETIC AND NATURAL FUELS** (Any fuel materials other than those defined under other categories, e.g., coal, nuclear fuels, hydrogen, etc.)
090100	**Hydrocarbon Fuels** (Methane, ethane, SNG, fuel oils, and gases when produced by chemical synthesis or by waste processing. Coverage includes production, chemical and physical properties, transport, economics, environmental aspects, and safety.)
090110	*Properties* (Chemical, physical, and thermodynamic properties of interest to fuel specialists.)
090120	*Preparation*
090121	Chemical Synthesis
090122	Preparation from Wastes or Biomass (Biochemical or thermochemical conversion of agricultural, industrial, or municipal wastes or biomass for hydrocarbon-fuel production. Waste-processing systems for fuel production such as Purox Pyrolysis Process. Includes methane recovery from sanitary landfills and hydrocarbon recovery from latex-bearing plants. Secondary category is **SOLAR ENERGY,** *Biomass Production and Conversion*. See also **BIOMEDICAL SCIENCES, BASIC STUDIES**.)
090200	**Alcohol Fuels** (Methanol or other fuel alcohols. Coverage includes production, chemical and physical properties, transport, economics, environmental aspects, and safety.)
090210	*Properties* (Chemical, physical, and thermodynamic properties of interest to fuel specialists.)
090220	*Preparation*
090221	Chemical Synthesis

Category No. **Category Title**

090222 Preparation from Wastes or Biomass
(Biochemical or thermochemical conversion of agricultural, industrial, or municipal wastes or biomass for alcohol-fuel production. Includes pretreatment processes (hydrolysis, heat treatments, etc.). Secondary category is **SOLAR ENERGY,** ***Biomass Production and Conversion.*** See also **BIOMEDICAL SCIENCES, BASIC STUDIES.**)

090300 **Inorganic Hydrogen Compound Fuels**
(Includes ammonia, hydrazine, and hydrogen peroxide, but excludes metallic hydrides. Covers production, chemical and physical properties, transport, economics, environmental aspects, and safety.)

090310 *Properties*
(Chemical, physical, and thermodynamic properties of interest to fuel specialists.)

090320 *Preparation*

090400 **Solid Waste and Wood Fuels**
(Only the direct use of waste, including biomass waste, or other solids such as wood or cellulose as a fuel; use and preparation of refuse-derived solid fuels and wood charcoal are included. If the waste is processed to produce a liquid or gaseous fuel, categorize as appropriate in **Hydrocarbon Fuels** or **Alcohol Fuels**. Secondary category is **SOLAR ENERGY,** ***Biomass Production and Conversion.***)

090500 **Liquid Waste Fuels**
(Only the direct use of waste as a fuel; if the waste is processed to produce a fuel, categorize as appropriate in **Hydrocarbon Fuels** or **Alcohol Fuels.**)

090600 **Gaseous Waste Fuels**
(Only the direct use of waste as a fuel; if the waste is processed to produce a fuel, categorize as appropriate in **Hydrocarbon Fuels** or **Alcohol Fuels.**)

130000 HYDRO ENERGY
(Low-head and small-scale hydroelectric power plants, retrofitting existing dams for power, hydroelectric-dam safety and environmental studies, and generating equipment suitable for small-scale and low-head plants. Also the extraction of energy from the Florida Current, Gulf Stream, or undammed, free-flowing streams.)

130100 **Resources and Availability**
(Resource assessments, reservoir management studies for hydroelectric power plants, dam operation, flow-rate studies for streams, and watershed runoff.)

130200 **Site Geology and Meteorology**
(Soil stability, seismicity, dam-foundation studies, rainfall, and snow-melt studies.)

130300 **Plant Design and Operation**
(Overall design only; components are included under **Power-Conversion Systems.**)

130400 **Regulations and Licensing**

130500 **Economics and Management**
(Economic feasibility studies. For reservoir management, see **Resources and Availability.**)

130600 **Environmental Aspects**
(Dam safety studies and fish-passage facilities, as well as usual environmental studies.)

130601 *Land Use and Aesthetics*

130700 **Power-Conversion Systems**
(Water-current generators, turbine designs, etc.)

140000 SOLAR ENERGY
(Conversion of solar radiation to useful amounts of electric energy, the use of solar energy for heating and cooling, or any other use of solar energy that might contribute to the total energy budget. All technical aspects of the design, research and development, manufacture, testing, and operation of solar cells and solar collectors are included. Also includes materials with indicated utility in solar cells or solar converters.)

140100 **Resources and Availability**
(Solar flux and insolation data, models, and measuring instruments.)

140300 **Economics**

140400 **Environmental, Legal, and Institutional Aspects**

Category No.	Category Title
140500	**Solar Energy Conversion**
140501	*Photovoltaic Conversion* (Research and development technology on solar cells and solar cell arrays, i.e., design, performance, theory, materials, optimization, and processing and production studies. See also subcategory **Photovoltaic Power Systems.**)
140502	*Thermionic and Thermoelectric Conversion* (Thermionic and thermoelectric conversion using a solar heat source.)
140504	*Biomass Production and Conversion* (Selection, growth, and harvest of terrestrial plants for energy; availability and collection of forest, agricultural, and animal residues; anaerobic digestion and fermentation of biomass to produce gaseous and liquid fuels; thermochemical conversion of biomass including gasification, liquefaction, and direct combustion; production of chemical feedstocks and electricity from biomass. See also subcategories under **BIOMEDICAL SCIENCES, BASIC STUDIES,** and under **OTHER SYNTHETIC AND NATURAL FUELS.**)
140505	*Photochemical, Photobiological, and Thermochemical Conversion* (Basic and applied research on modified photosynthetic processes, biomimetic processes, photochemical and photoelectrochemical reactions, and thermochemical and thermoelectrochemical reactions for conversion of solar energy to chemicals, fuels, electricity, and stored energy. See also subcategories under **CHEMISTRY; BIOMEDICAL SCIENCES, BASIC STUDIES;** and **HYDROGEN.**)
140600	**Photovoltaic Power Systems** (Small-scale, intermediate-size, and large-scale photovoltaic power systems for residential, community, and utility power production and remote applications such as irrigation, telecommunications, and spacecraft power supplies. For information on basic research and development of solar cells and solar-cell arrays, see the subcategory *Photovoltaic Conversion.*)
140700	**Solar Thermal Power Systems** (Systems for converting solar thermal energy to electrical or mechanical power. Includes small thermal power applications for communities, commercial/industrial users, irrigation systems, etc.; subsystem components such as heat engines, solar receivers, etc.; and total-energy and hybrid systems. Also includes large-scale systems (10–300 MWe) for electric power generation (new or retrofit).)
140702	*Central Receiver*
140703	*Distributed Collector*
140704	*Total Energy and Hybrid Systems* (Systems for the production of electricity and the utilization of rejected heat for such purposes as space heating and cooling for communities. Also, the utilization of complementary solar thermal-electric systems in conjunction with fossil-fuel and other energy systems.)
140800	**Ocean Energy Systems** (All aspects of ocean thermal energy conversion (OTEC) systems based on exploitation of the temperature difference between the surface water and ocean depth. All aspects of salinity-gradient power systems based on extracting energy from mixing of fresh water with seawater or a more-saline solution, using methods such as reverse electrodialysis, osmosis, or vapor-pressure differential. For ocean-current conversion systems, see **HYDRO ENERGY.** For wave-energy conversion, see **TIDAL AND WAVE POWER, Wave Energy Converters.** For nonconvecting or salinity-gradient solar-pond power systems, see **Solar Thermal Power Systems.**)
140900	**Solar Thermal Utilization**
140901	*Space Heating and Cooling*
140905	*Agricultural and Industrial Process Heat* (Crop drying, food dehydration, etc., and process heat for industrial purposes.)
140907	*Water Heating* (Service hot water for use in residential and commercial buildings.)
140909	*Miscellaneous Solar Applications* (Solar cookers, desalinators, refrigerators, etc.)
141000	**Solar Collectors and Concentrators** (Design and performance and collector optics, materials, sizing, etc.)
142000	**Heat Storage** (Sensible, latent, and thermochemical heat storage.)

Category No.	Category Title
150000	**GEOTHERMAL ENERGY**
150100	**Resource Status and Assessment** (Estimated reserves, current and predicted use, and rate of development and distribution of hydrothermal convective systems, conductive thermal gradients, geopressured systems, and magma and volcanic systems.)
150101	*USA*
150102	*Non-USA*
150200	**Geology and Hydrology of Geothermal Systems** (Tectonic and stratigraphic setting of the area; recent faulting; distribution and age of young volcanic rocks; location and character of thermal manifestations, including hydrothermally altered rock; temperature and discharge measurements of hot and cold springs; and well water table. Also includes mathematical and geological modeling.)
150201	*USA*
150202	*Non-USA*
150300	**Geothermal Exploration and Exploration Technology**
150301	*Geophysical Techniques and Surveys* (Temperature or geothermal-gradient surveys, heat-flow determinations, electric and electromagnetic surveys, seismic methods, and gravimetric surveys.)
150302	*Geochemical Techniques and Surveys* (Chemical analysis of thermal water and gases, silica content, isotope ratios, salinity, etc.)
150303	*Exploratory Drilling and Well Logging* (Temperature-depth distribution, pressure-depth distribution, permeability, porosity, lithology and stratigraphy, and fluid composition. Also includes downhole instrumentation.)
150400	**Legal and Institutional Aspects** (Effects of federal, state, and local laws and regulations on geothermal development; societal implications; taxation policies; and land-use criteria.)
150500	**Economic and Financial Aspects**
150600	**Environmental Aspects and Waste Disposal** (Ground subsidence, noise, earthquakes, uncontrolled blowouts, gaseous emissions, surface- and groundwater effects, and emission control and abatement.)
150700	**By-Products** (Chemical and mineral extraction, etc.)
150800	**Geothermal Power Plants** (Commercial pilot and demonstration plants and closed thermal loops.)
150801	*Design and Operation*
150802	*Power Plant Systems and Components* (Advanced conversion systems, heat exchangers, pumps, etc.)
150900	**Geothermal Engineering**
150901	*Drilling Technology and Well Hardware* (Advanced drill systems and methods of drilling, drilling fluids, cementation, downhole drill motors, etc.)
150902	*Fluid Transmission* (Steam and hot water transmission, two-phase transmission, and pipeline design and construction.)
150903	*Corrosion, Scaling, and Materials Development*
150904	*Geothermal Reservoir and Well Performance* (Well measurements—pressure, temperature, flow, etc.)
150905	*Safety and Control*
150906	*Reservoir Stimulation and Extraction Technology* (Nuclear-explosive stimulation, chemical-explosive stimulation, hydraulic fracturing, etc.)
151000	**Direct Energy Utilization** (Space heating and cooling, heat production for desalination and agriculture, etc.)
152000	**Geothermal Data and Theory** (Basic geoscience studies relating to geothermal energy development.)
152001	*Properties of Aqueous Solutions* (Thermodynamic, thermophysical, and electric properties; solute species identification; viscosity; and analytical chemistry of natural systems.)
152002	*Properties of Minerals and Rocks* (Zeolites, micas, chlorites, clays, amphibole, pyroxenes, hydrogrossular, gibbsite, glasses, etc.

Category No.	Category Title
	Includes thermal conductivity, electric properties, porosity, permeability, thermodynamic properties, and analytical chemistry of natural materials.)
152003	*Rock-Water-Gas Interactions* (Reaction products of water on minerals; solubility studies; equilibrium studies; hydrothermal alteration and synthesis; dissolution and precipitation of silica, silicates, calcite, and anhydrite from aqueous solutions.)
152004	*Isotope and Trace Element Studies* (Distribution coefficients and chemical and isotopic geothermometry.)
160000	**TIDAL AND WAVE POWER**
160300	**Regulations** (Legislation, hearings, government policy statements, etc.)
160400	**Economics** (Cost estimates, comparative cost studies, etc.)
160500	**Environmental Aspects** (Site environmental studies and environmental-impact studies in construction and operation of tidal power plants.)
160700	**Tidal Power Plants** (General studies on plants for converting tidal power to electric power.)
160701	*Design and Operation* (Design, development, construction, operation, and performance.)
160702	*Power Conversion Systems* (Turbogenerator design, operation, and performance.)
160703	*Site Characteristics* (Power potential and suitability of specific sites.)
160800	**Wave Energy Converters** (Wave power potential and wave energy-conversion devices.)
170000	**WIND ENERGY** (For information on wind wave conversion, see **TIDAL AND WAVE POWER, Wave Energy Converters.**)
170100	**Availability (Climatology)** (Regional or national wind-power or wind-velocity distributions in annual or seasonal averages. Information on specific land areas for use as wind-turbine sites is included under *Site Characteristics*.)
170300	**Regulations** (All legal aspects related to the use of wind energy for direct or indirect energy conversion.)
170400	**Economics** (Wind-turbine design and construction costs and economics of operating wind power-conversion systems.)
170500	**Environmental Aspects** (Effects of wind turbines on the environment and the effects of site construction, power-conversion systems, and power transmission from the site.)
170600	**Wind Energy Engineering**
170601	*Applications* (Diverse uses of wind energy-conversion systems.)
170602	*Turbine Design and Operation* (Theoretical and applied information on the machinery for wind energy-conversion systems.)
170603	*Power-Conversion Systems* (Conversion of wind-activated rotating or reciprocating shaft power to other forms of energy and also the direct use of wind pressure, e.g., electrofluiddynamic generators.)
170604	*Site Characteristics* (Data such as topography and meteorology on specific locations for wind energy-conversion systems.)
200000	**ELECTRIC POWER ENGINEERING** (Primarily reserved for DOE-, EPRI-, and TVA-sponsored research relating to utility-size fossil-fueled power plants. Routine aspects of power plant hardware are not included: examples

Category No.	Category Title
	of such hardware are fuses, motors, turbines, transformers, converters, generators, switchgear, cables, and standard electrical components. For nontechnological aspects of electric power, e.g., surveys and forecasting, utility management, rate-structure studies, power-consumption statistics, regulations, load management, and public relations, see **ENERGY PLANNING AND POLICY, Electric Power.**)
200100	**Power Plants and Power Generation**
	(Design, operation, and performance of fossil-fueled thermal power plants. See also **NUCLEAR POWER PLANTS, HYDRO ENERGY, SOLAR ENERGY,** and other energy-source-related categories.)
200101	*Cooling and Heat Transfer Equipment and Systems*
	(New hardware development and application.)
200102	*Power Cycles*
	(Use of improved cycles or combined cycles for increased efficiency or reliability.)
200103	*Waste-Fueled Systems*
	(Systems with refuse-derived fuels (RDF) or RDF/fossil fuels fired or co-fired in utility boilers. See also **OTHER SYNTHETIC AND NATURAL FUELS, Solid Waste and Wood Fuels** and **ENERGY CONSERVATION, CONSUMPTION, AND UTILIZATION, Waste Heat Recovery and Utilization.**)
200104	*Components*
	(Reliability, service life, and new or improved designs. See also **ENGINEERING.**)
200105	*Waste Heat Utilization*
	(Design and development of combined heat and power (CHP) plants. Also, direct use of hot water from fossil-fuel plant cooling systems for agricultural or other purposes. See also **ENERGY CONSERVATION, CONSUMPTION, AND UTILIZATION, Waste Heat Recovery and Utilization.**)
200106	*Economics*
	(Improvements in efficiency. Use of new technology, e.g., fluidized-bed combustion.)
200107	*Off-Peak Energy Storage*
	(See also **ENERGY STORAGE.**)
200108	*Fuels*
	(In-plant fuel preparation and handling. Plant conversion from one fuel to another. See also **COAL AND COAL PRODUCTS** and **PETROLEUM.**)
200200	**Environmental Control Technology**
	(Limited to on-site effluent control equipment and environmental protection measures. For off-site environmental aspects, see the **ENVIRONMENTAL SCIENCES** categories.)
200201	*Thermal Effluents*
200202	*Noxious Gas and Particle Emissions*
200203	*Site Selection and Land Use*
200300	**Power Transmission and Distribution**
	(Design studies or equipment developments for improved performance or reliability.)
200301	*AC Systems, EHV and UHV*
	(EHV—345 to 765 kV; UHV—over 800 kV.)
200302	*DC Systems*
200303	*Superconducting and Cryogenic Systems*
210000	NUCLEAR POWER PLANTS
	(Design, development, construction, and operation of fission power reactors. If indicated, the reactor type category should be primary except for: (1) process heat reactors, (2) control systems, (3) environmental aspects, or (4) reactor safety. Aspects of fuels mining, enrichment, processing, reprocessing, and waste management are categorized under **NUCLEAR FUELS.**)
210100	**Power Reactors, Non-Breeding, Light-Water Moderated, Boiling Water Cooled**
	(BWR-type reactors and reactor plants as energy sources for electricity generation and other applications—design, construction, performance, and operation, including fuel elements, components, and accessories, but not control systems, environmental aspects, or safety.)
210200	**Power Reactors, Non-Breeding, Light-Water Moderated, Non-Boiling Water Cooled**
	(PWR, WWER, etc., types. Reactors and reactor plants as energy sources for electricity generation and other applications—design, construction, performance, and operation, including fuel elements, components, and accessories, but not control systems, environmental aspects, or safety.)

Category No.	Category Title
210300	**Power Reactors, Non-Breeding, Graphite Moderated** (GCR, AGR, HTGR, Molten Salt, etc., types. Reactors and reactor plants as energy sources for electricity generation and other applications—design, construction, performance, and operation, including fuel elements, components, and accessories, but not control systems, environmental aspects, or safety.)
210400	**Power Reactors, Non-Breeding, Otherwise Moderated or Unmoderated** (Heavy-water-type reactors. Reactors and reactor plants as energy sources for electricity generation and other applications—design, construction, performance, and operation, including fuel elements, components, and accessories, but not control systems, environmental aspects, or safety.)
210500	**Power Reactors, Breeding** (LMFBR, MSBR, GCFR, and LWBR types. Reactors and reactor plants as energy sources for electricity generation and other applications—design, construction, performance, and operation, including fuel elements, components, and accessories, but not control systems, environmental aspects, or safety.)
210600	**Power Reactors, Auxiliary, Mobile, Package, and Transportable** (SNAP type, military nonpropulsion, etc. Reactors and reactor plants as energy sources for electricity generation and other applications—design, construction, performance, and operation, including fuel elements, components, and accessories, but not control systems, environmental aspects, or safety.)
210700	**Regulation and Licensing** (General licensing procedures and regulations, inspection, and legal aspects. Accident liability is included under **NUCLEAR REACTOR TECHNOLOGY, Reactor Safety**.)
210800	**Economics** (General economic studies. For economic aspects of specific reactors, use the primary category for reactor type.)
210801	*Construction and Operation* (Financing, capital costs, decommissioning costs, operating costs not associated with the fuel cycle, personnel requirements and training, etc.)
210802	*Fuel Cycle* (General aspects of the overall fuel cycle (mining to reprocessing), comparisons of cycles, studies of proliferation-resistant cycles, on-site storage of fresh and spent fuel, etc.)
210900	**Process Heat Reactors** (Reactors to provide heat for industrial processes or space heating, including coal gasification, hydrogen production, smelting, desalination, and district heating. Such reactors are assigned a secondary category by type, e.g., HTGR, if indicated.)
220000	NUCLEAR REACTOR TECHNOLOGY (General studies of reactor physics and engineering, environmental aspects, and nonpower reactors.)
220100	**Theory and Calculation** (Reactor physics calculations and in-pile measurements. For studies on physics of specific reactors, use reactor type category only. Includes criticality, dynamics, kinetics, lattice cell parameters, neutronics, physics, and thermohydraulics.)
220200	**Components and Accessories** (Design, construction, assembly, and performance when no reactor type is specified. If reactor type is specified, only appropriate type category is used. Includes buildings, coolants, cooling systems, loading machines, pressure vessels, shielding, and structures.)
220300	**Fuel Elements** (Design, assembly, and performance of fuel elements and assemblies, fuel-handling procedures, and fuel requirements when no reactor type is specified. If reactor type is specified, only appropriate type category is used. Excludes production and fabrication of fuel-element component materials, fuel-cycle economics, and loading machines.)
220400	**Control Systems** (Systems for the control and surveillance of reactors of any type. Secondary categories are used to indicate reactor type. Includes control elements and incorporated instruments, control-element drive units, control by computers, data-processing and display systems, alarm and automatic-shutdown systems, automatic initiation of protective systems or actions, and control theory. Excludes instruments of potential use in control systems and safety and protective systems, e.g., ECCS.)

Category No.	Category Title
220500	**Environmental Aspects** (Secondary category is to reactor type, if applicable. Effluent aspects are included only for generation, on-site treatment, and release. Off-site effluent monitoring, transport, and effects are included under the **ENVIRONMENTAL SCIENCES** categories.)
220501	*Siting* (Selection criteria and suitability studies.)
220502	*Radioactive Effluents* (Generation, on-site treatment, and release only.)
220503	*Chemical and Thermal Effluents* (Generation, on-site treatment, and release only.)
220600	**Research, Test, and Experimental Reactors** (Zero-power and training reactors and subcritical assemblies. Design, construction, performance, and operation, including fuel elements, components, and accessories, but not control systems, environmental aspects, or safety.)
220700	**Plutonium and Isotope Production Reactors** (Production reactors (producing fissionable materials), irradiation reactors such as chemonuclear reactors, isotope production reactors, tritium production reactors, materials testing reactors, and materials processing reactors—design, construction, performance, and operation, including fuel elements, components, accessories, and associated plants, but not control systems, environmental aspects, or safety.)
220800	**Propulsion Reactors** (Includes aircraft, rocket, and ship reactors.)
220801	*Nuclear Ship Licensing and Regulation*
220900	**Reactor Safety** (Accident analysis, safety system performance, radiation protection, decommissioning and dismantling, and site and plant security measures. Secondary categories are used to indicate reactor type.)
220901	*Accident Liability* (Legal aspects of nuclear damage and risk, insurance, and financial security.)
250000	ENERGY STORAGE (Methods for storing energy in a readily recoverable form for later use. Such methods may be mechanical, chemical, electromagnetic, or thermal.)
250100	**Magnetic**
250200	**Compressed Gas**
250300	**Pumped Hydro** (See also **HYDRO ENERGY**.)
250400	**Capacitor Banks**
250500	**Flywheels** (For information on the use of flywheels for vehicle propulsion, see **ADVANCED AUTOMOTIVE PROPULSION SYSTEMS, Flywheel Propulsion**.)
250600	**Thermal**
250700	**Liquefied Gas**
250800	**Chemical**
250900	**Batteries** (Chemical batteries of the large-current, rechargeable type. All technical aspects are included, e.g., electrodes, electrolytes, components, charging, gas emission, manufacture, testing, reliability, standards, and maintenance. Materials with indicated usage in battery construction are also covered. For applications of such batteries as power sources for land vehicles or electric power storage by utilities for use during peak consumption, see **ADVANCED AUTOMOTIVE PROPULSION SYSTEMS, Electric-Powered Systems** and **ELECTRIC POWER ENGINEERING,** *Off-Peak Energy Storage*, respectively. Information on so-called solar batteries or nuclear batteries is excluded; such information is included under **SOLAR ENERGY** and **ISOTOPE AND RADIATION SOURCE TECHNOLOGY**.)
250901	*Design and Development*
250902	*Performance and Testing*

Category No.	Category Title
250903	*Materials, Components, and Auxiliaries*
250904	*Other Applications*
290000	**ENERGY PLANNING AND POLICY** (Information dealing with all aspects of energy planning, policy, and policy analysis. Includes information in the following areas: energy policy, organization, regulations, and legislation; conservation and consumption of energy resources; supply and demand; economic evaluations and cost comparisons; shortages, including blackouts and brownouts; long- and short-term planning; research and development in new technology; total energy systems; district heating and cooling; combined cycles in energy management; energy transport, e.g., pipelines and liquid hydrogen. Additional categories should be used where appropriate.)
290100	**Energy Analysis and Modeling** (Methodology or techniques for process analysis, input-output analysis, systems analysis, regional analysis, etc., including energy accounting and net energy evaluations. For analysis for a specific purpose, see the other subcategories; e.g., for forecasting electric power demand, see the subcategory **Electric Power**.)
290200	**Economics and Sociology** (All economic, sociological, or political aspects of any energy action or policy, e.g., energy source development, energy facility construction and/or siting, implementation of conservation or other programs, the OPEC cartel, etc.; energy tax and tax-credit programs; energy prices; and impacts on households of energy expenditures. Also includes basic economic and social studies that are applicable to analysis of energy systems or policy. See also **ENVIRONMENTAL–SOCIAL ASPECTS OF ENERGY TECHNOLOGIES, Social and Economic Studies**.)
290300	**Environment, Health, and Safety** (Planning and policy aspects as opposed to technological studies found in the **ENVIRONMENTAL SCIENCES** categories; energy or monetary costs of complying with regulations; environmental impacts of energy source development, energy facility construction, and/or siting; conventional energy supplies, etc.; hazardous materials and wastes, including acid precipitation; environmental regulations and legislation per se. See also **ENVIRONMENTAL–SOCIAL ASPECTS OF ENERGY TECHNOLOGIES**.)
290400	**Natural Resources** (Broad survey-type or economics-oriented studies related to resource development or utilization. Includes ores, minerals, and materials generally, but not specifically, unless an element of unusual interest, e.g., uranium. Also includes general studies on water resources, land use, food production, and resource conservation.)
290500	**Research, Development, Demonstration, and Commercialization** (Overall energy R, D, and D programs of federal, state, or local governments or organizations; industrial laboratories; regional organizations; private institutions; and foreign governments and organizations. Technology assessment, transfer, and utilization are also included.)
290600	**Nuclear Energy** (Only the broader nontechnological aspects of nuclear energy as related to overall energy policy or strategy, including nuclear weapons proliferation issues or nonproliferation policies; also the social, institutional, and public policy aspects of the nuclear power controversy. For technological information, see the categories **NUCLEAR FUELS; NUCLEAR POWER PLANTS;** and **NUCLEAR REACTOR TECHNOLOGY**.)
290700	**Transport and Storage** (Energy transport or energy storage, not energy aspects of transportation or energy-supply stockpiling or storage. For technical aspects, see **ENERGY STORAGE**.)
290800	**Waste Heat Utilization** (Utilitarian or public policy aspects or feasibility studies of total energy systems, cogeneration, district heating and cooling, or any other use of waste heat. For technical aspects, see **ELECTRIC POWER ENGINEERING,** *Waste Heat Utilization* and **ENERGY CONSERVATION, CONSUMPTION, AND UTILIZATION,** *Waste Heat Recovery and Utilization*.)
291000	**Conservation** (Policy, legislative, and regulatory aspects of conservation, including energy management and energy auditing; also nontechnological studies of efficient use of existing resources, e.g., underground space, or of innovative construction, e.g., earth-sheltered buildings. For technical

Category No.	Category Title
	information in the specific consuming sectors, see the category **ENERGY CONSERVATION, CONSUMPTION, AND UTILIZATION.**)
292000	**Supply, Demand, and Forecasting** (Surveys, analyses, simulations, forecasts, and forecasting techniques for predicting energy demand, supply, or impacts; emergency measures, e.g., allocations, substitutions, etc., to insure adequate supplies; impacts of weather on supply and demand.)
293000	**Policy, Legislation, and Regulation** (Congressional hearings, legislation, regulations, etc., for implementing or interpreting already-established policy or that can aid in formulating policy; assessments, evaluations, surveys, regional hearings, etc., on present or to-be-established policies. Information concerning energy policies of state, local, and foreign governments or regional or international organizations is also included.)
294000	**Fossil Fuels** (Policies and programs for and economic, sociological, and environmental aspects of coal, petroleum, natural gas, and oil shale or tar sand development; also all assessments of reserves and resources, both U. S. and worldwide. For technological studies, see the specific fossil-fuel main categories, e.g., **COAL AND COAL PRODUCTS; PETROLEUM;** etc.)
294001	*Coal*
294002	*Petroleum*
294003	*Natural Gas*
294004	*Oil Shales and Tar Sands*
295000	**Hydrogen and Synthetic Fuels** (Assessment of the potential of hydrogen in a hydrogen-based economy or of nonfossil fuel sources in substitution for conventional sources. See also the categories **HYDROGEN** and **OTHER SYNTHETIC AND NATURAL FUELS.**)
296000	**Electric Power** (Surveys and forecasting of electricity demand and supply; statistics of power consumption on a local, regional, or national basis; utility management problems including planning, rate structures and regulations, accounting procedures, peak-load pricing, load management, power thefts, public relations, etc. For technical aspects of electric power generation, transmission, or distribution, see **ELECTRIC POWER ENGINEERING.**)
296001	*Electric Power Generation*
296002	*Electric Power Transmission and Distribution*
298000	**Consumption and Utilization** (Survey- or statistical-type information on energy usage in the consuming sectors. For information of a more technical nature, see the category **ENERGY CONSERVATION, CONSUMPTION, AND UTILIZATION.**)
299000	**Unconventional Sources and Power Generation** (Policy and management aspects of solar, geothermal, wind, tidal, wave, or other renewable energy sources; for technical aspects, see the main categories for these energy sources.)
299001	*Solar*
299002	*Geothermal*
299003	*Other* (Winds, tides, wood, plants, wastes, etc.)
300000	**DIRECT ENERGY CONVERSION** (Methods and devices for converting heat or other forms of energy into electrical energy without intermediate conversion into mechanical work. See also **ISOTOPE AND RADIATION SOURCE TECHNOLOGY, Isotopic Power Supplies; SOLAR ENERGY,** *Photovoltaic Conversion;* and **SOLAR ENERGY,** *Thermionic and Thermoelectric Conversion.*)
300100	**MHD Generators** (For theoretical studies of magnetohydrodynamics not necessarily applicable to MHD generators, see **PHYSICS I, Fluid Physics,** *Magnetohydrodynamics.*)
300101	*Design and Development*
300102	*Performance and Testing*

Category No.	Category Title
300103	*Materials, Components, and Auxiliaries*
300104	*Duct Engineering and Fluid Dynamics*
300200	**EHD Generators** (For theoretical studies of electrohydrodynamics not necessarily applicable to EHD generators, see **PHYSICS I, Fluid Physics,** *Electrohydrodynamics.*)
300300	**Thermoelectric Generators** (A device that converts heat energy directly into electric energy by using the Seebeck effect. See also **ISOTOPE AND RADIATION SOURCE TECHNOLOGY, Isotopic Power Supplies** and **SOLAR ENERGY,** *Thermionic and Thermoelectric Conversion.*)
300301	*Design and Development*
300302	*Performance and Testing*
300303	*Materials, Components, and Auxiliaries*
300304	*Applications*
300400	**Thermionic Converters** (See also **ISOTOPE AND RADIATION SOURCE TECHNOLOGY, Isotopic Power Supplies** and **SOLAR ENERGY,** *Thermionic and Thermoelectric Conversion.*)
300401	*Design and Development*
300402	*Performance and Testing*
300403	*Materials, Components, and Auxiliaries*
300404	*Applications*
300500	**Fuel Cells** (All technical aspects of regenerative and nonregenerative fuel cells are included either as a complete unit or when various components or parameters are evaluated, such as fuels, electrolytes, electrodes, power output, and efficiency. Economic and commercial appraisals of fuel cells for use in vehicles, power stations, space power systems, etc., are also included.)
300501	*Design and Development*
300502	*Performance and Testing*
300503	*Materials, Components, and Auxiliaries*
300504	*Applications*
300505	*Electrochemistry, Mass Transfer, and Thermodynamics*
300800	**Miscellaneous Converters** (Piezoelectric, ferroelectric, ferromagnetic, Nernst, magnetothermoelectric, photoelectromagnetic, magnetostrictive conversion, etc.)
320000	**ENERGY CONSERVATION, CONSUMPTION, AND UTILIZATION** (Equipment and methods to reduce energy consumption, increase energy efficiency, or enable the substitution of more-plentiful energy sources for those in short supply. Development and properties of specific fuels or energy sources are included at the specific fuel or energy source categories, e.g., **COAL AND COAL PRODUCTS; PETROLEUM; NATURAL GAS;** etc. Engineering aspects of electric power generation and transmission are included under **ELECTRIC POWER ENGINEERING.** Management and policy information is included under **ENERGY PLANNING AND POLICY.**)
320100	**Buildings** (Research, development, and demonstration activities related to energy conservation in buildings and consumer products, e.g., the development and commercialization of energy-efficient building materials, methods, and processes for retrofitting existing buildings as well as in new construction. Includes information on more-efficient heating and cooling systems, e.g., heat pumps, lighting systems, appliances, and thermal insulation, and on fuel and/or electric power consumption in buildings. Also includes the use of landscaping, winterization, architecture, etc., to reduce energy consumption.)
320101	*Residential Buildings* (Single and multiple units.)
320102	*Office Buildings*
320103	*School, Municipal, and Other Public Buildings* (Hospitals, nursing homes, prisons, detention homes, etc.)

Category No.	Category Title
320104	*Commercial and Industrial Buildings* (Food stores, banks, repair shops, restaurants, department stores, shopping centers, manufacturing plants, etc.)
320200	**Transportation** (Measures to improve flow of high-occupancy vehicles, improve total vehicle flow, increase car and van occupancy, improve the efficiency of taxi service, and reduce the need for travel; transportation pricing; fuel consumption; etc. Includes the use of more energy-efficient transportation systems and measures to improve fuel economy, such as better operating and maintenance procedures. For actual engineering and design studies of more energy-efficient power plants and vehicles, see **ADVANCED AUTOMOTIVE PROPULSION SYSTEMS**.)
320201	*Air and Aerospace*
320202	*Railway*
320203	*Land and Roadway*
320204	*Sea and Water*
320205	*Pipeline*
320300	**Industry and Agriculture** (Information on reducing energy consumption or waste, increasing industrial process efficiency, optimizing materials, increasing agricultural productivity and food-processing efficiency, etc.)
320301	*Energy Sources* (Evaluations of alternative energy sources or fuel substitution, fuel and/or power consumption data, improved methods for fuel conversion, energy requirements for industrial processes, impact of energy shortages on agricultural or industrial production, etc. See also **ENERGY PLANNING AND POLICY**.)
320302	*Materials* (Optimization of materials for reducing energy consumption and the flow of materials within and among industrial process streams.)
320303	*Equipment and Processes* (Design, evaluation, maintenance, optimization, or replacement of industrial or agricultural equipment and processes to achieve meaningful energy savings.)
320304	*Waste Heat Recovery and Utilization* (Recovery and use of waste heat, systems analyses for reducing waste heat, development of bottoming cycles, cogeneration systems, etc. For waste heat utilization by electric power plants, see **ELECTRIC POWER ENGINEERING,** ***Waste Heat Utilization***.)
320305	*Industrial Waste Management* (All areas of industrial waste processing for energy or resource recovery, including economic studies, environmental effects, government policies, and product acceptability. For the use of wastes as fuels or processing to obtain fuels, see **ELECTRIC POWER ENGINEERING** or **OTHER SYNTHETIC AND NATURAL FUELS.** See also **ENGINEERING, Waste Processing Plants and Equipment**. For radioactive waste management, see **NUCLEAR FUELS, Waste Management**.)
320600	**Municipalities and Community Systems** (Integrated community energy systems (ICES), district heating and cooling systems, land use and planning, mixed energy systems, etc.)
320601	*Conservation Services* (Institutional barriers to energy conservation, e.g., financial restrictions, and recommendations for overcoming such constraints, including voluntary measures, energy-management surveys, municipal building codes and standards, and zoning and land-use planning for energy conservation.)
320602	*Public Services* (Design, operation, and maintenance of street-lighting systems, roads and streets, and recreational facilities such as swimming pools and tennis courts for maximum energy conservation. For mass transit, see **Transportation**.)
320603	*Public Utilities* (Energy-conservation measures and policy and legislation affecting municipal power systems, water-treatment plants and distribution systems, natural gas distribution systems, and sewer systems. Includes district heating and cooling systems, integrated community energy systems (ICES), modular integrated utility systems (MIUS), effects of utility pricing on energy conservation, utility fuel supplies, demand for and charges for fuel or power services, etc. For

Category No.	Category Title
	design, operation, and performance of fossil-fuel power plants and power transmission, see **ELECTRIC POWER ENGINEERING.**)
320604	*Municipal Waste Management*
	(Commercial and municipal waste and sewage processing for energy recovery or recycling. For information on facilities for collecting, sorting, or processing solid wastes, see the category **ENGINEERING, Waste Processing Plants and Equipment.**)
320900	**Education and Public Relations**
	(Consumer motivation and behavior, school curricula, educational tools, public opinion polls, training aids, etc., that deal with energy conservation and consumption.)
330000	**ADVANCED AUTOMOTIVE PROPULSION SYSTEMS**
	(Design and development of advanced propulsion systems for automobiles, buses, trucks, ships, and trains—e.g., power plants, components, and devices that promise better fuel economy, less maintenance, and increased service life; more-efficient power cycles; better emission-control devices; and feasibility studies on the use of alternative fuels, such as hydrogen or alcohol fuels.)
330100	**Internal Combustion Engines**
330101	*Spark-Ignition*
330102	*Diesel*
330103	*Turbine*
330104	*Rotary*
330200	**External Combustion Engines**
330201	*Stirling Cycle*
330202	*Rankine Cycle*
330300	**Electric-Powered Systems**
330400	**Hybrid Systems**
330500	**Flywheel Propulsion**
330600	**Vehicle Design Factors**
330601	*Body and Chassis*
330602	*Engine-Transmission Matching*
330603	*Engine System*
330604	*Weight*
330700	**Emission Control**
	(Hardware only; for environmental aspects, see the **ENVIRONMENTAL SCIENCES** categories.)
330701	*Nitrogen Oxides*
330702	*Carbon Monoxide*
330703	*Sulfur Dioxide*
330704	*Hydrocarbons*
330705	*Particulates*
330800	**Alternative Fuels**
	(Refer also to specific categories for alternative fuels, such as **NATURAL GAS; HYDROGEN;** and **OTHER SYNTHETIC AND NATURAL FUELS.**)
360000	**MATERIALS**
	(Any material directly associated with energy production, conversion, or utilization. Radiation effects or radiation damage to any material is included with the following exceptions: radiation effects on biological materials are categorized under **BIOMEDICAL SCIENCES, APPLIED STUDIES**; radiation effects on chemical or physicochemical properties are categorized under **CHEMISTRY**. Ion channeling in materials is included when the radiation-effects aspects are considered; studies on the particle-transport aspects are categorized under **PHYSICS II, Radiation and Shielding Physics,** and studies on ion-surface or ion-atom collision aspects are categorized under **PHYSICS I, Atomic, Molecular, and Chemical Physics.** Materials used as fuel, e.g., coal, uranium, and thorium, are categorized under the appropriate fuel category.

Category No.	Category Title
	Studies on destructive and nondestructive testing are included under **ENGINEERING, Materials Testing**.)
360100	**Metals and Alloys** (Specifically Ac, Am, At, B, Be, Bk, Cd, Ce, Cf, Cm, Dy, Er, Es, Eu, Fm, Fr, Gd, Hf, Ho, In, La, Lr, Lu, Md, Mo, Nb, Nd, No, Np, Pa, Pm, Po, Pr, Pu, Ra, Re, Ru, Sc, Sm, Ta, Tb, Th, Tm, U, V, W, Y, Yb, and Zr. Also includes any metal or alloy associated with an actual or simulated energy-related application. Aspects of thorium, plutonium, and uranium used as a nuclear fuel are categorized under **NUCLEAR FUELS; NUCLEAR POWER PLANTS;** or **NUCLEAR REACTOR TECHNOLOGY**.)
360101	*Preparation and Fabrication* (Bonding, brazing, casting, cold working, drawing, extrusion, fastening, forging, forming, hot working, molding, pressing, rolling, sintering, soldering, swaging, welding, etc.)
360102	*Structure and Phase Studies* (Allotropy, crystal-phase transformations, melting points, microstructure, phase diagrams, solidification, etc.)
360103	*Mechanical Properties* (Brittleness, buckling, cracking, creep, deformation, ductility, elastic properties, elongation, embrittlement, fatigue, fracture, friction, hardness, plasticity, Poisson's ratio, rupture, shear properties, strain, strength, stress, tensile properties, toughness, wear, Young's modulus, etc.)
360104	*Physical Properties* (Damping, density, electrical properties, internal friction, magnetic properties, optical properties, specific heat, thermal conductivity, thermal diffusivity, thermal expansion, thermodynamic properties, transformation temperature, vapor pressure, superconductivity, etc.)
360105	*Corrosion and Erosion* (Includes oxidation, hydridation, and sulfidation.)
360106	*Radiation Effects* (All radiation effects on mechanical integrity or physical properties of metals or alloys. Radiation effects on chemical properties is categorized under **CHEMISTRY, Radiation Chemistry**.)
360200	**Ceramics, Cermets, and Refractories** (Specifically the borides, carbides, hydrides, nitrides, and oxides of Ac, Am, At, B, Be, Bk, Cd, Ce, Cf, Cm, Dy, Er, Es, Eu, Fm, Fr, Gd, Hf, Ho, In, La, Lr, Lu, Md, Mo, Nb, Nd, No, Np, Pa, Pm, Po, Pr, Pu, Ra, Re, Ru, Sc, Sm, Ta, Tb, Th, Tm, U, V, W, Y, Yb, and Zr or any metal associated with an energy-related application.)
360201	*Preparation and Fabrication* (Bonding, brazing, casting, cold working, drawing, extrusion, fastening, forging, forming, hot working, molding, pressing, rolling, sintering, soldering, swaging, welding, etc.)
360202	*Structure and Phase Studies* (Allotropy, crystal-phase transformations, melting points, microstructure, phase diagrams, solidification, etc.)
360203	*Mechanical Properties* (Brittleness, buckling, cracking, creep, deformation, ductility, elastic properties, elongation, embrittlement, fatigue, fracture, friction, hardness, plasticity, Poisson's ratio, rupture, shear properties, strain, strength, stress, tensile properties, toughness, wear, Young's modulus, etc.)
360204	*Physical Properties* (Damping, density, electrical properties, internal friction, magnetic properties, optical properties, specific heat, thermal conductivity, thermal diffusivity, thermal expansion, thermodynamic properties, transformation temperature, vapor pressure, superconductivity, etc.)
360205	*Corrosion and Erosion* (Includes oxidation, hydridation, and sulfidation.)
360206	*Radiation Effects* (All radiation effects on mechanical integrity or physical properties of ceramics, cermets, and refractories.)
360300	**Composite Materials** (Actual or simulated energy-related applications of combinations of alloys or plastics, usually with the addition of strengthening agents.)
360301	*Preparation and Fabrication* (Bonding, brazing, casting, cold working, drawing, extrusion, fastening, forging, forming, hot working, molding, pressing, rolling, sintering, soldering, swaging, welding, etc.)

Category No.	Category Title
360302	*Structure and Phase Studies* (Allotropy, crystal-phase transformations, melting points, microstructure, phase diagrams, solidification, etc.)
360303	*Mechanical Properties* (Brittleness, buckling, cracking, creep, deformation, ductility, elastic properties, elongation, embrittlement, fatigue, fracture, friction, hardness, plasticity, Poisson's ratio, rupture, shear properties, strain, strength, stress, tensile properties, toughness, wear, Young's modulus, etc.)
360304	*Physical Properties* (Damping, density, electrical properties, internal friction, magnetic properties, optical properties, specific heat, thermal conductivity, thermal diffusivity, thermal expansion, thermodynamic properties, transformation temperature, vapor pressure, superconductivity, etc.)
360305	*Corrosion, Erosion, and Degradation* (Includes oxidation, hydridation, and sulfidation.)
360306	*Radiation Effects* (All radiation effects on mechanical integrity or physical properties of composite materials.)
360400	**Polymers and Plastics** (Actual or simulated energy-related applications; for radioinduced polymerization, see **CHEMISTRY**.)
360401	*Preparation and Fabrication* (Bonding, brazing, casting, cold working, drawing, extrusion, fastening, forging, forming, hot working, molding, pressing, rolling, sintering, soldering, swaging, welding, etc.)
360402	*Structure and Phase Studies* (Allotropy, crystal-phase transformations, melting points, microstructure, phase diagrams, solidification, etc.)
360403	*Mechanical Properties* (Brittleness, buckling, cracking, creep, deformation, ductility, elastic properties, elongation, embrittlement, fatigue, fracture, friction, hardness, plasticity, Poisson's ratio, rupture, shear properties, strain, strength, stress, tensile properties, toughness, wear, Young's modulus, etc.)
360404	*Physical Properties* (Damping, density, electrical properties, internal friction, magnetic properties, optical properties, specific heat, thermal conductivity, thermal diffusivity, thermal expansion, thermodynamic properties, transformation temperature, vapor pressure, superconductivity, etc.)
360405	*Degradation and Erosion* (Includes oxidation, hydridation, and sulfidation.)
360406	*Radiation Effects* (All radiation effects on mechanical integrity or physical properties of polymers and plastics. For effects on chemical reactions, including polymerization, see **CHEMISTRY.**)
360600	**Other Materials** (Specifically energy-related applications of boron, carbon, graphite, and concretes. Additional examples are glass, semiconductor materials, soil, rock, cloth, etc. See also **GEOSCIENCES** for information on soil and rock mechanics, etc.)
360601	*Preparation and Manufacture* (Bonding, brazing, casting, cold working, drawing, extrusion, fastening, forging, forming, hot working, molding, pressing, rolling, sintering, soldering, swaging, welding, etc.)
360602	*Structure and Phase Studies* (Allotropy, crystal-phase transformations, melting points, microstructure, phase diagrams, solidification, etc.)
360603	*Properties* (Brittleness, buckling, cracking, creep, deformation, ductility, elastic properties, elongation, embrittlement, fatigue, fracture, friction, hardness, plasticity, Poisson's ratio, rupture, shear properties, strain, strength, stress, tensile properties, toughness, wear, Young's modulus, etc. Also includes damping, density, electrical properties, internal friction, magnetic properties, optical properties, specific heat, thermal conductivity, thermal diffusivity, thermal expansion, thermodynamic properties, transformation temperature, vapor pressure, superconductivity, etc.)
360604	*Corrosion, Erosion, and Degradation* (Includes oxidation, hydridation, and sulfidation.)
360605	*Radiation Effects* (Radiation effects on mechanical integrity and physical properties of these materials.)

Category No.	Category Title
400000	**CHEMISTRY**
400100	**Analytical and Separations Chemistry** (Analytical chemistry of isotopes, transactinides, actinides, lanthanides, Be, Cs, I, Kr, Mo, Nb, Pm, Rn, Ra, Ru, Sc, Sr, Ta, Tc, Te, Xe, Y, Zr, or any element or compound directly associated with energy production, utilization, or conversion. Analyses to determine reactor fuel burnup; burnup calculations are included under **NUCLEAR REACTOR TECHNOLOGY.** Arrangement is according to analytical procedure. Information relative to analytical methods is included as indicated.)
400101	*Activation and Nuclear Reaction Procedures* (The subject in general or descriptions of the method irrespective of the materials involved.)
400102	*Chemical Procedures* (Limited to chemical analytical procedures for fuels, energy-related materials, and the specific elements listed above.)
400103	*Radiometric and Radiochemical Procedures* (Nuclear radiometric or radiochemical procedures, including properties of radioisotopes as sources in x-ray analyses irrespective of the materials involved. Routine tracer studies are excluded.)
400104	*Spectral Procedures* (Limited to spectral analytical procedures for fuels, energy-related materials, and the specific elements listed above. For structural studies using spectral procedures, see **Inorganic and Physical Chemistry; Organic Chemistry;** or **MATERIALS,** *Structure and Phase Studies.*)
400105	*Separation Procedures* (Separation procedures for transactinides, actinides, lanthanides, Be, Cs, I, Kr, Mo, Nb, Pm, Rn, Ra, Ru, Sc, Sr, Ta, Tc, Te, Xe, Y, Zr, and any material of interest in energy production, utilization, and conversion, except the procedures covered elsewhere, e.g., under *Activation and Nuclear Reaction Procedures;* **NUCLEAR FUELS;** and **COAL AND COAL PRODUCTS.**)
400200	**Inorganic and Physical Chemistry**
400201	*Chemical and Physicochemical Properties* (Chemical and physicochemical properties of nonradioactive elements, inorganic compounds, or inorganic mixtures directly associated with energy production, utilization, or conversion. See also **GEOTHERMAL ENERGY, Geothermal Data and Theory.**)
400202	*Isotope Effects* (Isotope effects on nonnuclear chemical and physical properties of elements and inorganic compounds. Isotope effects are not generally included when used only as a tool in the analysis of reaction mechanisms or in structure studies. See also **Organic Chemistry.**)
400203	*Isotope Exchange and Isotope Separation* (Isotope exchange between elements and inorganic compounds if the exchange is of primary concern or the exchange mechanism is used in isotope separation. See also **Organic Chemistry.** Chemical and physicochemical isotope separations are included here. Preparation of labeled inorganic compounds using stable isotopes is also included here if there is a contribution to the technology. Processes for uranium enrichment are included under **NUCLEAR FUELS, Enrichment;** processes for other production-scale processes are included under **ISOTOPE AND RADIATION SOURCE TECHNOLOGY.** Radioisotope production is included under **Radiochemistry and Nuclear Chemistry.**)
400300	**Organic Chemistry**
400301	*Chemical and Physicochemical Properties* (Chemical and physicochemical properties of organic compounds and mixtures directly associated with energy production, utilization, or conversion. Includes preparation of radioprotective compounds.)
400302	*Isotope Effects* (Isotope effects on nonnuclear chemical and physical properties of organic compounds. Isotope effects are not generally included when used only as a tool in the analysis of reaction mechanisms or structure studies. See also **Inorganic and Physical Chemistry.**)
400303	*Isotope Exchange and Isotope Separation* (See note under same subheading at **Inorganic and Physical Chemistry.**)
400400	**Electrochemistry** (Chemical reactions induced by electricity in any form or accompanied by electric phenomena if the reactions are directly related to energy production, utilization, or conversion.)

Category No.	Category Title

400500 **Photochemistry**

(Chemical reactions and electronic excitation of molecules induced by light when directly related to energy production, utilization, or conversion or when included in comparison studies with radiation chemistry.)

400600 **Radiation Chemistry**

(The radiations considered here include accelerated ions, gamma radiation, x radiation, bremsstrahlung, neutrons, protons, alpha particles, beta particles, deuterons, and other subatomic particles. Ultraviolet, infrared, and visible radiations are not included. Radiation effects on the chemical and physicochemical properties of any element, compound, or mixture. General papers on solvated electrons and hydroxyl radicals are also included. Radiation effects on the biological activity of biochemicals in vitro or in vivo and all information on radiation effects on biochemicals in living organisms is included under **BIOMEDICAL SCIENCES, APPLIED STUDIES.** Radiation effects on metallurgical or solid-state properties are included under **MATERIALS.**)

400700 **Radiochemistry and Nuclear Chemistry**

(Radiochemistry and nuclear chemistry are included without restrictions except for tracer applications and labeled compound preparation, which are included only if there is a contribution to the technology. See also **NUCLEAR FUELS** and **PHYSICS II, Nuclear Physics.**)

400701 *Hot-Atom Chemistry*

(Any aspect, including recoil production and reactions.)

400702 *Properties of Radioactive Materials*

(Chemical and physicochemical properties of radioactive elements or materials, including the stability of labeled compounds and chemical decontamination of surfaces or work areas. Decontamination of reactor vessels is included under **NUCLEAR REACTOR TECHNOLOGY.** Decay properties of specific nuclides are included under **PHYSICS II, Nuclear Physics.**)

400703 *Radioisotope Production*

(Radioisotope production processes involving chemical separations. Development and uses of radioisotope radiation sources or power sources are included under **ISOTOPE AND RADIATION SOURCE TECHNOLOGY.**)

400800 **Combustion, Pyrolysis, and High-Temperature Chemistry**

(Basic research on combustion, pyrolysis, and high-temperature reactions when not covered by specific categories such as **COAL AND COAL PRODUCTS, Combustion; EXPLOSIONS AND EXPLOSIVES, Chemical;** etc. Includes theory of combustion, flame mechanisms and kinetics, flame structure and properties, detonations and explosions in combustion, energy and mass transfer in combustion, and charged particles in combustion. Also includes mechanisms and kinetics of reactions induced by high temperature as in pyrolysis or occurring at high temperature as in coal-fired MHD generators. Fires, such as forest or structural, are excluded. Research on combustion equipment is categorized under **ENGINEERING, Combustion Systems.** Research on MHD systems and components is categorized under **DIRECT ENERGY CONVERSION, MHD Generators.**)

420000 ENGINEERING

(Interest is restricted to the specific subject concepts indicated in the following subcategories. In general, these are energy-related engineering aspects that are not restricted to a single energy source or the specific energy source is not explicitly indicated.)

420100 **General Engineering**

(General engineering studies for which a more appropriate category has not been provided. Restricted to DOE-sponsored work only.)

420200 **Facilities and Equipment**

(Design, construction, operation, and applied-mechanics studies (stress, strain, motion, vibration, elasticity, deformation, etc.) of facilities and equipment including components of such facilities and equipment, e.g., irradiation plants, pilot plants, laboratories, etc. In general, information on facilities and equipment is included under categories more closely related to the purpose of the facility or equipment, e.g., **CHEMISTRY; PARTICLE ACCELERATORS; NUCLEAR POWER PLANTS;** etc. Interest is restricted to DOE-sponsored work only.)

Category No. **Category Title**

420201 *Cryogenic and Superconducting Equipment and Devices*
(Design, fabrication, performance, and testing of superconducting equipment and devices, e.g., superconducting cables, magnets, generators, etc., or equipment and devices using superconductors or superconducting junctions as a component, e.g., superconducting quantum interference devices (SQUIDs), Josephson junctions, etc. Cryogenic equipment and devices are also included when such equipment and devices are used in the production of low temperatures or in studies of superconductivity. Applications of superconducting devices are included under categories for the application, e.g., **PARTICLE ACCELERATORS; FUSION ENERGY; ELECTRIC POWER ENGINEERING,** etc. Information on the theory of superfluidity is included under **PHYSICS I, Fluid Physics.** Information on the theory of superconductivity is included under **PHYSICS II, Solid State Physics.** Information on superconducting materials is included under **MATERIALS.**)

420202 *Protective Structures and Equipment*
(Structures and equipment designed primarily for equipment and personnel protection, e.g., blast and fallout shelters, shielding—except for accelerators and reactors, clean rooms, air filtration systems, fire protection systems, masks, and special clothing. The fabrication of shielding materials is included under **MATERIALS;** shielding calculations are included under **PHYSICS II, Radiation and Shielding Physics.**)

420203 *Handling Equipment and Procedures*
(Design, construction, and operation of equipment and facilities for handling radioactive materials, e.g., remote-handling equipment, glove boxes, and hot cells. Nonreactor criticality studies are also included.)

420204 *Shipping Containers*
(Design, development, evaluation, safety analyses, and testing of radioactive-materials shipping containers.)

420205 *Transport and Storage Facilities*
(Design, development, evaluation, and testing of tanks, pipelines, tanker vessels, and other containers and their auxiliary equipment for transport and storage of nonradioactive energy-related materials. For studies of a fuel under conditions of transport and storage, see specific fuel categories.)

420206 *Mining and Drilling Equipment and Facilities*
(Equipment and facilities when no specific application is given; e.g., offshore platforms and drilling rigs. For coal mining equipment and facilities, see **COAL AND COAL PRODUCTS, Mining.**)

420300 **Lasers**
(Laser research and development information; not applications. Applications of lasers in isotope separation are included under **ISOTOPE AND RADIATION SOURCE TECHNOLOGY; NUCLEAR FUELS;** and **CHEMISTRY.** Applications of lasers in fusion plasma production and heating are included under **FUSION ENERGY.**)

420400 **Heat Transfer and Fluid Flow**
(Heat-transfer and fluid-flow studies only if nucleate boiling, boiling burnout, critical heat flux, two-phase flow, flow in rod bundles, or liquid metals are considered or if DOE-sponsored work is not adequately categorized elsewhere. Specifically excluded are heat exchangers except where DOE-sponsored.)

420500 **Materials Testing**
(Destructive or nondestructive materials testing if a direct application to energy research and development is indicated and if categorization on the basis of application would be less appropriate. In general, information on component testing is included under categories more closely related to the component, e.g., information on reactor fuel-element testing is included under **NUCLEAR REACTOR TECHNOLOGY.**)

420600 **Safety Engineering**
(Safety studies at energy research or production facilities unless a more appropriate category is provided. In general, safety engineering information is included under categories for the specific activities or facilities, e.g., **NUCLEAR REACTOR TECHNOLOGY** or **EXPLOSIONS AND EXPLOSIVES.** Refer also to **HEALTH AND SAFETY.**)

420700 **Vacuum Engineering**
(Vacuum engineering associated with energy research or production facilities.)

420800 **Electronic Circuits and Devices**
(Design, development, performance, and testing of electronic circuits and devices related to

Category No.	Category Title
	energy research. More appropriate categories are used when available, e.g., information on instruments or circuits for instruments is categorized under **INSTRUMENTATION.**)
420900	**Waste Processing Plants and Equipment**
	(Engineering of facilities and equipment for collecting, sorting, or processing municipal or other wastes only if for conversion to fuel or for recycle. Actual conversion to synthetic fuels or direct use as fuel is included under **OTHER SYNTHETIC AND NATURAL FUELS** or **SOLAR ENERGY, Solar Energy Conversion,** *Biomass Production and Conversion.* Disposal of radioactive waste is included under **NUCLEAR FUELS, Waste Management.** Routine disposal of municipal and industrial wastes without conversion to fuel or recycle is specifically excluded.)
421000	**Combustion Systems**
	(Design and development of combustion systems such as boilers, furnaces, etc., and their components only where improvements are considered that increase efficiency or decrease pollution or if existing designs are adapted to new fuels or fuel mixtures. Chemical aspects of combustion are included under **CHEMISTRY, Combustion, Pyrolysis, and High-Temperature Chemistry.** Combustion of coal, petroleum or petroleum products, natural gas, hydrogen, or biomass is included under other appropriate categories. Combustion equipment used in vehicles should be included in the category **ADVANCED AUTOMOTIVE PROPULSION SYSTEMS.**)
422000	**Underground Engineering**
	(Equipment, facilities, and methods, including economic and feasibility studies, for construction of underground structures, e.g., cavities, drill holes, mine shafts, tunnels, etc., only when related specifically to the production, storage, utilization, or conversion of energy and not adequately categorized elsewhere. Drilling information specifically related to petroleum and natural gas production, coal mining, and geothermal exploration is included under their corresponding categories. Information on the peaceful use of nuclear explosions for the construction of cavities, reservoirs, craters, canals, etc., is included under **EXPLOSIONS AND EXPLOSIVES.** Water wells and mineral mines are specifically excluded.)
423000	**Marine Engineering**
	(Design, construction, and operation of equipment for offshore operations only when specifically related to energy production, storage, conversion, or utilization and not adequately categorized elsewhere. Information on specific offshore energy sources such as OTEC, tidal and wave power, offshore oil wells, etc., is included under their corresponding categories.)
424000	**Pollution Control Equipment**
	(Equipment and methods for controlling the emission of pollutants from industrial plants. Information on pollution-control equipment and methods for thermal power plants is included under **ELECTRIC POWER ENGINEERING,** and that for vehicles is included under **ADVANCED AUTOMOTIVE PROPULSION SYSTEMS.** Routine waste treatment is specifically excluded.)
425000	**Power Cycles**
	(General studies not appropriately categorized elsewhere. Power cycles used in power plants or vehicles are found in the corresponding categories.)
425001	*Brayton*
425002	*Rankine*
425003	*Stirling*
425007	*Other*
430000	**PARTICLE ACCELERATORS**
430100	**Design, Development, and Operation**
	(Information related to accelerators in general. Specific research facilities are included under *Experimental Facilities and Equipment.*)
430200	**Beam Dynamics, Field Calculations, and Ion Optics**
	(Information on beam dynamics in fields. See also *Injection and Extraction Systems* and *Experimental Facilities and Equipment.*)
430300	**Auxiliaries and Components**
430301	*Ion Sources*
	(Design, development, operation, and performance of ion sources for accelerator applications only. Information on ion sources without specific applications is included under **PHYSICS I, Atomic, Molecular, and Chemical Physics,** *Beams and Their Reactions.*)

Category No.	Category Title
430302	*Injection and Extraction Systems* (Beam-transport systems are included under *Experimental Facilities and Equipment.*)
430303	*Experimental Facilities and Equipment* (Beam-transport systems, target facilities, and other research facilities.)
430400	**Storage Rings** (Design, development, and operation of particle storage rings.)
440000	INSTRUMENTATION (Radiation instrumentation and radiation effects on instrumentation. Information on other instrumentation is included in categories more closely related to the application of the instrument. In retrieval, the user should always use a combined category and subject search procedure.)
440100	**Radiation Instrumentation** (All aspects of nuclear instrumentation; excludes visible radiation (light), radiofrequency, microwave, or infrared instrumentation. Accessory equipment is categorized to the appropriate category for the instrument, e.g., amplifiers for gamma spectrometers go to *Nuclear Spectroscopic Instrumentation.*)
440101	*General Detectors and Monitors*
440102	*Radiation Dosemeters*
440103	*Nuclear Spectroscopic Instrumentation* (Instruments for measurement of energy spectra of nuclear particles or radiation, e.g., gamma spectrometers, beta spectrometers, etc.)
440104	*High Energy Physics Instrumentation* (Detectors for tracking subnuclear particles or radiation, e.g., bubble chambers, Cherenkov counters, gas track detectors, missing-mass spectrometers, and spark chambers.)
440105	*Radiometric Instruments* (Instruments containing both a source and detector usually employed in dimensional or material-property measurements, e.g., thickness gages, moisture gages. Excludes radiometers.)
440200	**Radiation Effects on Instrument Components, Instruments, or Electronic Systems**
440300	**Miscellaneous Instruments** (Miscellaneous instruments associated with energy research. Examples are EPR, NMR, optical, and mass spectrometers; infrared, microwave, radiofrequency, and ultraviolet detectors; and chemical-process instruments.)
440400	**Well Logging Instrumentation** (Instrumentation and procedures for specific applications are categorized under COAL AND COAL PRODUCTS, **Reserves and Exploration**; PETROLEUM, **Geology and Exploration**; NATURAL GAS, **Geology and Exploration**; OIL SHALES AND TAR SANDS, **Reserves and Exploration**; NUCLEAR FUELS, **Exploration**; and GEOTHERMAL ENERGY, *Exploratory Drilling and Well Logging.*)
450000	EXPLOSIONS AND EXPLOSIVES
450100	**Chemical** (DOE-sponsored studies on chemical or high explosives and explosions, such as preparation or production, analysis, initiation or detonation properties, physical or chemical properties, performance, safety, storage, etc. In addition, all other information on the use of chemical explosions to simulate or detonate nuclear explosions is included.)
450200	**Nuclear**
450201	*Civil Uses* (Engineering/construction, e.g., studies on the excavation of craters, cavities, and canals; the use of nuclear explosions for stimulating natural gas and petroleum production; ground motion and seismological studies related to civil uses; and the effects of nuclear explosions on facilities and equipment.)
450202	*Weaponry* (Testing of nuclear devices, including the legal or political effects of such testing; nonproliferation, safeguarding, and inspection of nuclear weapons; and military uses, including the deployment of nuclear weapons and the effects of such explosions on facilities and equipment.)

Category No.	Category Title
450300	**Explosion Detection** (All studies related to the detection of nuclear explosions, e.g., information on the discrimination between nuclear explosions and natural phenomena, such as earthquakes.)
500000	ENVIRONMENTAL SCIENCES, ATMOSPHERIC
500100	**Basic Studies** (Any basic atmospheric studies that establish information from which environmental effects of energy-related activities can be predicted, calculated, or measured.)
500101	*Radiometric Techniques* (Use of radioisotopes or ionizing radiations in basic atmospheric investigations.)
500200	**Chemicals Monitoring and Transport** (Transport, monitoring, characterization, removal, or reactions of pollutants—e.g., carbon, lead, nitrogen, or sulfur compounds, or respirable particulates—from the energy cycle during transit through or presence in the atmosphere; includes all atmospheric aspects of acid precipitation. Also includes measures for control and abatement of these pollutants. Biological effects of these pollutants or chemicals are included under **BIOMEDICAL SCIENCES, APPLIED STUDIES.**)
500300	**Radioactive Materials Monitoring and Transport** (Atmospheric studies on radioactive materials and radioisotopes or ionizing radiations, especially from nuclear power plants or any phase of the fuel cycle. Atmospheric consequences of nuclear weapons testing is included, e.g., fallout information. See **NUCLEAR FUELS, Environmental Aspects.**)
500400	**Thermal Effluents Monitoring and Transport** (Studies of atmospheric consequences of moisture and heat releases from power plant cooling towers or from any energy-related technology. Includes studies on cooling tower plume behavior, aerodynamic and thermal exchange over cooling ponds, and meteorological and climate studies. Also includes effects from energy-conservation measures, such as thermal pollution from increased use of heat pumps. Also includes control and abatement. Biological effects of thermal effluents or elevated temperatures are included under **BIOMEDICAL SCIENCES, APPLIED STUDIES.**)
500500	**Site Resource and Use Studies** (Meteorological and atmospheric aspects of existing or potential sites for any phase of energy development and use.)
500600	**Regulations** (Legislation, ordinances, rules, litigation, hearings, decisions, etc. Recommended or adopted guides, criteria, and standards for allowable atmospheric emission limits or pollutant concentrations; air-quality standards. Any item categorized here should be cross-categorized to **ENERGY PLANNING AND POLICY, Environment, Health, and Safety.**)
510000	ENVIRONMENTAL SCIENCES, TERRESTRIAL
510100	**Basic Studies** (Any basic geological or soil studies that establish information from which environmental effects of energy-related activities can be predicted, calculated, or measured.)
510101	*Radiometric Techniques* (Use of radioisotopes or ionizing radiation in basic terrestrial studies.)
510200	**Chemicals Monitoring and Transport** (Transport, monitoring, characterization, removal, or reactions of pollutants—e.g., carbon, lead, nitrogen, or sulfur compounds—including hazardous wastes, from the energy cycle during transit through or presence in the terrestrial environment or ecosystem; includes all terrestrial aspects of acid precipitation. Also includes measures for control and abatement of these pollutants. Biological effects of these pollutants or chemicals are included under **BIOMEDICAL SCIENCES, APPLIED STUDIES.**)
510300	**Radioactive Materials Monitoring and Transport** (Terrestrial studies on radioactive materials and radioisotopes or ionizing radiations, especially from nuclear power plants or any phase of the fuel cycle. Includes radionuclide migration, reactions, activity levels, and doses during transit through or presence in the terrestrial environment. See also **NUCLEAR FUELS, Environmental Aspects.**)
510301	*Soil*

Category No.	Category Title
510302	*Terrestrial Ecosystems and Food Chains*
510400	**Thermal Effluents Monitoring and Transport** (Studies on impact of heat releases on soils, terrestrial ecosystems, rocks, or underground openings. Measures for control and abatement are included. Biological effects of thermal effluents or elevated temperatures are included under **BIOMEDICAL SCIENCES, APPLIED STUDIES.**)
510500	**Site Resource and Use Studies** (Terrestrial and geological aspects of existing or potential sites for any phase of energy development and use. Includes land use and reclamation.)
510600	**Regulations** (Legislation, ordinances, rules, litigation, hearings, decisions, etc. Recommended or adopted guides, criteria, and standards for allowable terrestrial emission limits or pollutant concentrations, or land use and reclamation. Any item categorized here should be cross-categorized to **ENERGY PLANNING AND POLICY, Environment, Health, and Safety.**)
520000	ENVIRONMENTAL SCIENCES, AQUATIC
520100	**Basic Studies** (Any basic studies of aquatic systems and ecology that contribute information from which environmental effects of energy-related activities can be predicted, calculated, or measured.)
520101	*Radiometric Techniques* (Use of radioisotopes or ionizing radiation in basic aquatic studies.)
520200	**Chemicals Monitoring and Transport** (Transport, monitoring, characterization, removal, or reactions of pollutants—e.g., carbon, lead, nitrogen, or sulfur compounds—including hazardous wastes, from the energy cycle during transit through or presence in groundwater, surface waters, or aquatic systems; includes surface water acidification resulting from acid precipitation. Also includes measures for control and abatement of these pollutants. Biological effects of these pollutants or chemicals are included under **BIOMEDICAL SCIENCES, APPLIED STUDIES.**)
520300	**Radioactive Materials Monitoring and Transport** (Aquatic studies on radioactive materials and radioisotopes or ionizing radiations, especially from nuclear power plants or any phase of the fuel cycle. Includes radionuclide migration, reactions, activity levels, and doses during transit through or presence in the aquatic environment. See also **NUCLEAR FUELS, Environmental Aspects.**)
520301	*Water* (Studies related to transport and behavior of radioactive materials in groundwater and surface waters.)
520302	*Aquatic Ecosystems and Food Chains* (Studies related to radionuclide behavior in aquatic ecosystems and food chains.)
520400	**Thermal Effluents Monitoring and Transport** (Studies of consequences and effects of the release of heated effluents into the aquatic environment. Includes studies of plume behavior, hydrodynamics of heated effluent mixing with the receiving water, and modeling of any related aspects. Measures for control and abatement are included. Biological effects of thermal effluents or elevated temperatures are included under **BIOMEDICAL SCIENCES, APPLIED STUDIES.**)
520500	**Site Resource and Use Studies** (Hydrological or aquatic ecological considerations in site evaluation or selection for any energy- or DOE-related activity.)
520600	**Regulations** (Legislation, ordinances, rules, litigation, hearings, decisions, etc. Recommended or adopted guides, criteria, and standards for allowable emission limits or pollutant concentrations in the aquatic environment; includes water-quality standards. Any item categorized here should be cross-categorized to **ENERGY PLANNING AND POLICY, Environment, Health, and Safety.**)
530000	ENVIRONMENTAL–SOCIAL ASPECTS OF ENERGY TECHNOLOGIES
530100	**Social and Economic Studies** (Studies of the effects of energy-related activities on the socioeconomic environment.)

Category No.	Category Title
530200	**Assessment of Energy Technologies** (Studies of cost/risk/benefit tradeoff, alternatives, integrated assessments, etc., with emphasis on environmental aspects.)
530300	**Environmental Impact Statements** (Final statements only.)
550000	**BIOMEDICAL SCIENCES, BASIC STUDIES** (Studies of living organisms and components of living organisms; identification of functions, activities, and phenomena associated with these organisms; and studies to establish norms from which effects of energy production, conversion, or utilization activities can be determined. This broad category and the subdivisions that follow are intended primarily for publications on basic research sponsored by the Department of Energy. It should not be assumed that these broadly defined subject areas are otherwise within scope of the Energy Data Base.)
550100	**Behavioral Biology** (Sociology, psychology, anthropology, learning, population studies. See also the **ENVIRONMENTAL SCIENCES** categories.)
550101	*Tracer Techniques* (Use of radioisotopes in behavioral investigations.)
550200	**Biochemistry** (Chemical reactions and compounds occurring in living organisms, including biosynthesis, enzymes, hormones, carbohydrates, lipids, nucleic acids, purines, pyrimidines, proteins, peptides, amino acids, sterols, steroids, vitamins, etc.)
550201	*Tracer Techniques* (Use of radioisotopes in biochemistry.)
550300	**Cytology** (Structure, behavior, growth, and reproduction of cells; functions and chemistry of cell constituents.)
550301	*Tracer Techniques* (Use of radioisotopes in cytology.)
550400	**Genetics** (Heredity and variation among related organisms; plant and animal breeding; genes; genetic variability; gene transfer.)
550401	*Tracer Techniques* (Use of radioisotopes in genetics.)
550500	**Metabolism** (Physical and chemical processes by which substances are transformed by cells and energy is made available for use by an organism; catabolism; assimilation; all forms of intake; tissue distribution.)
550501	*Tracer Techniques* (Use of radioisotopes in metabolic investigations.)
550600	**Medicine** (Prevention, cure, or alleviation of disease; dentistry; chemotherapy; preventive medicine; pharmacology; pediatrics; surgery; etc.)
550601	*Unsealed Radionuclides in Diagnostics* (Nonroutine diagnostic techniques; imaging procedures; non-imaging procedures in vivo; radioassay procedures in vitro (including radioimmunoassay); methods of incorporation and extraction of radionuclides and labeled compounds. Note: Routine uses of unsealed radionuclides in medical diagnosis are not included. For nonmedical aspects of internal radionuclides, see **BIOMEDICAL SCIENCES, APPLIED STUDIES,** *Nuclide Kinetics and Toxicology*.)
550602	*External Radiation in Diagnostics* (Nonroutine uses of sealed radionuclide sources, x rays, bremsstrahlung, neutrons, and charged particles in diagnostic techniques; imaging procedures. Note: Routine uses of radiations and nuclear techniques in diagnosis are not included.)
550603	*External Radiation in Therapy* (Nonroutine uses of sealed radionuclide sources, x rays, bremsstrahlung, neutrons, and charged particles in therapeutic procedures; surface and depth dose distributions. Note: Routine uses of external radiation in therapy are not included. For nonmedical uses of radiation, see **ISOTOPE**

Category No.	Category Title
	AND RADIATION SOURCE TECHNOLOGY. For side effects of radiotherapy, see **BIOMEDICAL SCIENCES, APPLIED STUDIES,** ***Radiation Effects on Animals.***)
550604	*Unsealed Radionuclides in Therapy* (Nonroutine therapeutic techniques; internal dose distributions; response-modifying factors; use of artificially induced activation for therapeutic purposes; methods of incorporation and extraction of radionuclides and labeled compounds. Note: Routine uses of unsealed radionuclides in therapy are not included.)
550700	**Microbiology** (All aspects of microorganisms, including bacteria, viruses, fungi, algae, protozoans, yeasts, rickettsiae, etc.)
550701	*Tracer Techniques* (Use of radioisotopes in microbiology.)
550800	**Morphology** (Form and structure of animals and plants; anatomy; morphogenesis.)
550801	*Tracer Techniques* (Use of radioisotopes in morphological investigations.)
550900	**Pathology** (Causes, nature, and conditions produced by diseases, aberrations, or other abnormalities including allergy, ascites, atrophy, biological shock, edema, fibrosis, fistulae, granulomas, hemolysis, hemorrhage, hypertrophy, inflammation, jaundice, leukopenia, malformations, necrosis, splenomegaly, and ulcers.)
550901	*Tracer Techniques* (Use of radioisotopes in pathological investigations.)
551000	**Physiological Systems** (Processes, activities, and functions of organs and parts of living systems, including endocrine, neural, reproductive, gastrointestinal, and cardiovascular systems; splanchnology; myology; hematology.)
551001	*Tracer Techniques* (Use of radioisotopes in physiological investigations.)
552000	**Public Health**
552001	*Tracer Techniques* (Use of radioisotopes in public health investigations.)
553000	**Agriculture and Food Technology** (Crops, food technology; animal husbandry; pest control; horticulture; growth regulation; fertilizers; agricultural economics; forestry; etc.)
553001	*Tracer Techniques* (Use of radioisotopes in investigations of agriculture and food technology.)
560000	**BIOMEDICAL SCIENCES, APPLIED STUDIES**
560100	**Radiation Effects** (For this category the term radiation is defined as nuclear particles, accelerated electrons and ions, and gamma and x radiations. Information on UV effects is included only if microorganisms, cells, or biochemicals are the subjects of study. Laser radiation, magnetic fields, microwaves, and sunlight are excluded. The concept of radiation effects includes the effect itself, radiomimetic effects, and any information relating to the effect, e.g., information on assessment, prevention, alteration, transmission, reversal, recovery, etc., and information on the determination or influence of experimental variables such as dose, dose rate, LET, etc. This includes modification of radiation effects by radioprotective or effect-enhancing substances and studies on these substances to establish their independent effects; these studies should also be categorized under **Chemicals Metabolism and Toxicology.** Theoretical aspects of dose determination or measurement in tissue-equivalent materials or phantoms are included under **PHYSICS II, Medical Physics.** Monitoring of radioactive materials in the environment is included under the **ENVIRONMENTAL SCIENCES** categories.)
560110	*Radiation Effects on Biochemicals* (Effects on biochemicals in living organisms and on their biological activity; information on other effects on biochemicals outside living organisms is included under **CHEMISTRY.** Also includes the preparation of materials for irradiation and irradiation procedures. The preparation of the radiation sources is included under **CHEMISTRY** or **ISOTOPE AND RADIATION SOURCE TECHNOLOGY.**)

Category No.	Category Title
560111	In Vitro
560112	In Microorganisms
560113	In Plants
560114	In Animals
560115	In Man
560120	*Radiation Effects on Cells* (Effects of radiation from external or internal sources on cells in vitro, including cells in cultures; effects on in vivo cells is included in categories for entire organisms. Also includes information on the preparation of materials for irradiation and irradiation procedures. The preparation of the radiation sources is included under **CHEMISTRY** or **ISOTOPE AND RADIATION SOURCE TECHNOLOGY.**)
560121	External Source
560122	Internal Source
560130	*Radiation Effects on Microorganisms* (Effects on microorganisms, including bacteria, protozoa, unicellular algae, fungi, and yeasts. Uses of radiation in industrial processes for the radiopreservation of food, radiosterilization of medical supplies, radiation preparation of vaccines, etc., are included under **ISOTOPE AND RADIATION SOURCE TECHNOLOGY** unless effects on specific organisms are considered.)
560131	Basic Studies
560132	Food Preservation
560133	Radiosterilization of Medical Supplies
560134	Vaccine Preparation and Other Applications
560140	*Radiation Effects on Plants* (Effects on plants, including multicellular algae and in vivo studies on plant cells, of external-source radiations including the preparation of materials for irradiation and irradiation procedures. The preparation of the radiation sources is included under **CHEMISTRY** or **ISOTOPE AND RADIATION SOURCE TECHNOLOGY.** Industrial processes for radiopreservation or disinfestation of foods is under **ISOTOPE AND RADIATION SOURCE TECHNOLOGY.**)
560141	Basic Studies
560142	Applications
560150	*Radiation Effects on Animals*
560151	Man (Effects of external-source or implanted-source radiations on man or on human tissues or cells in vivo. Therapeutic irradiation is included when there is significant information on the effects on normal tissues, when dose evaluations are made for new radiation sources, e.g., pion beams, or when the method is of primary concern. Also includes health physics of external radiation.)
560152	Vertebrates (Effects of external-source or implanted-source radiations on vertebrates or on vertebrate tissues or cells in vivo, including the preparation of animals for irradiation and irradiation procedures. The preparation of radiation sources is included under **CHEMISTRY** or **ISOTOPE AND RADIATION SOURCE TECHNOLOGY.**)
560153	Invertebrates (Effects on invertebrates or on invertebrate tissues or cells in vivo, including the preparation of the organism for irradiation and irradiation procedures. The preparation of radiation sources is included under **CHEMISTRY** or **ISOTOPE AND RADIATION SOURCE TECHNOLOGY.** Industrial-scale processes for pest control and crop disinfestation are included under **ISOTOPE AND RADIATION SOURCE TECHNOLOGY** unless there is significant information on the effects of radiation on specific organisms.)
560154	Basic Studies
560155	Disinfestation
560156	Pest Control
560160	*Radionuclide Effects—Internal Source* (Absorbed or incorporated radioactive materials, not implanted sources, in any in vivo

Category No.	Category Title
	studies. Internal-source evaluation and effects, including experimental procedures. Preparation of labeled compounds for administration is included under **CHEMISTRY.** Also includes health physics of internal radiation.)
560161	Man
560162	Animals
560163	Plants
560164	Microorganisms
560170	*Nuclide Kinetics and Toxicology* (Kinetics and toxicology of deuterium and radionuclides in or on living organisms. Also includes contamination and decontamination, both internal and external, of living organisms. Levels and movements of radionuclides through man's environment are included under the **ENVIRONMENTAL SCIENCES** categories.)
560171	Man
560172	Animals
560173	Plants
560174	Microorganisms
560200	**Thermal Effects** (Effects on living organisms of thermal effluents from energy production, utilization, or conservation activities or other comparable thermal discharges. Includes effects of decreased temperature resulting from hydroelectric dams. Laboratory studies of temperature effects on living organisms are also included. Monitoring and behavior of thermal effluents are included under the **ENVIRONMENTAL SCIENCES** categories.)
560201	*Cells*
560202	*Microorganisms*
560203	*Plants*
560204	*Invertebrates*
560205	*Vertebrates*
560206	*Man*
560300	**Chemicals Metabolism and Toxicology** (Effects of any element or compound associated with an energy cycle, including resource extraction, conversion, utilization, and waste processing and disposal. For effects of radioactive effluents, see the **Radiation Effects** categories. Monitoring and behavior of chemical effluents in the environment are included under the **ENVIRONMENTAL SCIENCES** categories.)
560301	*Cells*
560302	*Microorganisms*
560303	*Plants*
560304	*Invertebrates*
560305	*Vertebrates*
560306	*Man*
560400	**Other Environmental Pollutant Effects** (Effects of noise produced in energy production, conversion, or utilization activities; hazards from power-transmission lines; laser and microwave radiation hazards; other health hazards from energy-related activities that are not covered in other categories.)
570000	**HEALTH AND SAFETY** (General aspects only. Health and safety information related to specific materials or facilities is included under categories for the materials or facilities.)
580000	**GEOSCIENCES** (Basic geoscience studies, excluding applied studies for which more appropriate categories are provided. This broad category and the subdivisions that follow are intended primarily for publications on basic research sponsored by the Department of Energy. It should not be assumed that these broadly defined subject areas are otherwise within scope of the Energy Data Base.)

Category No.	Category Title
580100	**Geology and Hydrology** (Studies on geologic formations, stratigraphy, geohydrology, etc. Also mathematical models of groundwater transport. Geochronology and fossils are also included unless more appropriately categorized under **ENVIRONMENTAL SCIENCES, TERRESTRIAL**.)
580200	**Geophysics** (Solid earth and hydrosphere only. Atmospheric and magnetospheric geophysics are included under **PHYSICS I, Atmospheric Physics.**)
580201	*Seismology and Tectonics* (Studies on earthquake prediction. Also includes seismic-wave propagation in reservoir rock; seismic monitoring of specific installations; and seismotectonics of volcanic and high heat-flow areas.)
580202	*Volcanology* (Basic studies involving volcanism; excludes descriptions of phenomena.)
580203	*Geophysical Survey Methods* (Seismic, electrical, and thermal methods. Other categories are provided for applications of geophysical methods to petroleum, natural gas, uranium, and geothermal prospecting.)
580300	**Mineralogy, Petrology, and Rock Mechanics** (Dynamic rock-fracturing studies, laboratory and theoretical studies of fracture mechanics, and mathematical models or rock behavior in response to mechanical stimuli such as explosions, fluid injection, etc. Includes DOE-sponsored studies on properties of rocks, soils, and minerals. See also specific minerals and rocks under **GEOTHERMAL ENERGY, Geothermal Data and Theory**.)
580400	**Geochemistry** (DOE-sponsored studies on the distribution, circulation, and amounts of chemical elements in minerals, ores, rocks, soils, water, and atmosphere. See also the **ENVIRONMENTAL SCIENCES** categories and **GEOTHERMAL ENERGY, Geothermal Data and Theory**.)
580500	**Oceanography** (Study of the ocean, particularly water currents, sediments, seafloor stability, topography, wave forces, etc. See also the **ENVIRONMENTAL SCIENCES** categories for biological and ecological aspects and **Geophysics** for aspects of tectonics and thermal processes. For applied oceanography see **PETROLEUM; NATURAL GAS; ENGINEERING, Marine Engineering**, etc.)
640000	**PHYSICS I** (Physics is divided into two parts to enable proper subcategorization. Part I includes **Astrophysics and Cosmology; Atmospheric Physics; Atomic, Molecular, and Chemical Physics; Fluid Physics;** and **High Energy Physics**.)
640100	**Astrophysics and Cosmology** (The subject scope of this category is limited to that specifically stated under the following subcategories.)
640101	*Cosmic Radiation* (Cosmic radiation, including exposure ages and induced radioactivity of extraterrestrial material, e.g., meteorites and lunar material. Cosmic-ray interactions with the atmosphere and with the geomagnetic field are included under **Atmospheric Physics.** Information on nuclear reactions or particle-particle interactions in which cosmic radiation is used as a source of incident particles is included under **High Energy Physics** or **PHYSICS II, Nuclear Physics,** depending on emphasis.)
640102	*Stars* (Evolution and nucleogenesis of stars, including star clusters, neutron stars, black holes, and other starlike objects. Celestial mechanics and routine astronomical observations, e.g., sky surveys, are excluded.)
640103	*Quasi-Stellar, Radio, and X-Ray Sources* (Theory and nuclear aspects of such objects.)
640104	*Solar Phenomena* (Sunspots and solar prominences, disturbances, wind, flares, radio bursts, magnetic fields, and structure; do not include spectra unless specifically related to the foregoing. Solar-wind interactions with the atmosphere or geomagnetic field are included under **Atmospheric Physics.**)

Category No.	Category Title
640105	*Galaxies* (Evolution of galaxies and identification of molecular species in space.)
640106	*Cosmology* (Evolution of the universe, gravitational collapse and detection of gravitational waves, and tests of general relativity. Gravitational-wave detectors are included. General relativity theory and field theories of gravitation are included under **PHYSICS II, Theoretical Physics.**)
640107	*Planetary Phenomena* (Natural radioactivity of extraterrestrial materials, evolution of the solar and planetary systems, and radiation belts. Van Allen belts are included under **Atmospheric Physics.**)
640200	**Atmospheric Physics** (Upper atmosphere only; interest is restricted to specific subject concepts indicated. Information on the biosphere is included under the **ENVIRONMENTAL SCIENCES** categories.)
640201	*Auroral and Ionospheric Phenomena* (Associated particle kinetics only.)
640202	*Effects of Nuclear Detonations* (Atmospheric effects only.)
640203	*Magnetospheric Phenomena* (Particle kinetics in the magnetosphere, geomagnetic field, and Van Allen belts and solar wind interactions with the atmosphere and geomagnetic field. Routine measurements of the field strength or theories of the origin of the geomagnetic field are excluded.)
640300	**Atomic, Molecular, and Chemical Physics** (Interest is restricted to specific subject concepts indicated under the following subcategories.)
640301	*Beams and Their Reactions* (Methods of production and techniques of using atomic, electron, ion, and molecular beams; electron and ion sources not developed for specific applications; and reactions of beams with materials in bulk, e.g., scattering by surfaces, secondary emission, and sputtering yields.)
640302	*Atomic and Molecular Properties* (Atomic properties and experimental and theoretical studies of helium, hydrogen, deuterium, tritium, the lanthanides, scandium, technetium, yttrium, and elements having $Z > 83$. Molecular properties of deuterium, helium, hydrogen, tritium, and compounds of technetium and of elements having $Z > 83$. Hyperfine structure and isotope shifts for all elements.)
640303	*Positronium, Muonium, and Muonic and Mesic Atoms and Molecules* (Properties of positronium, muonium, and muonic and mesic atoms and molecules. Exotic atoms, involving antiprotons, hyperons, etc., are included under **PHYSICS II, Nuclear Physics.** Basic mechanisms of positron annihilation in materials are included under **PHYSICS II, Radiation and Shielding Physics.** High-energy electron-positron annihilation interactions for studying elementary-particle production and properties are included under **High Energy Physics.** The use of positron annihilation as a tool for studying material properties is excluded.)
640304	*Collision Phenomena* (Experimental and theoretical studies of elastic scattering, charge exchange, electron capture, electron loss, electron attachment or detachment, excitation, deexcitation, excitation transfer, ionization, or dissociation in electron-ion, electron-atom, electron-molecule, ion-ion, ion-atom, ion-molecule, atom-atom, and atom-molecule collisions. Rearrangement collisions, chemical exchange, and collisions in confined gases when no external beam is involved are excluded.)
640305	*Atomic and Molecular Theory* (Theory of atomic and molecular structure and properties when generally applicable; also includes quantum properties of macromolecules.)
640400	**Fluid Physics** (Interest is limited to specific subject concepts indicated under the following subcategories.)
640410	*General Fluid Dynamics* (Theoretical treatments only; dynamics of engineering fluids are included under **ENGINEERING, Heat Transfer and Fluid Flow.** Interest is restricted to DOE-supported research.)
640420	*Properties and Structure of Fluids* (Low temperature plasma studies not related to fusion energy. Interest is restricted to DOE-supported research.)
640430	*Magnetohydrodynamics* (Theoretical treatments of conducting fluids in magnetic or crossed electric and magnetic fields;

Category No.	Category Title
	experimental studies applicable to MHD generators are included under **DIRECT ENERGY CONVERSION**.)
640440	*Electrohydrodynamics*
	(Theoretical treatments of dielectric fluids in electric fields; experimental studies applicable to EHD generators are included under **DIRECT ENERGY CONVERSION**.)
640450	*Superfluidity*
	(All experimental and theoretical aspects of superfluidity and superfluids; information on cryogenic fluids related to energy production, utilization, or conversion is included under categories for the application, e.g., **ELECTRIC POWER ENGINEERING, DIRECT ENERGY CONVERSION, ENGINEERING**, etc.)
640460	*Other Quantum Fluids*
645000	**High Energy Physics**
645100	*Particle Interactions and Properties—Experimental*
	(Experimental studies on the electromagnetic, weak, and strong interactions and on properties of particles and radiation quanta. Intermediate energy physics ($N\pi$ and Δ-1236 resonance dynamics in nuclear media, hypernuclei, and hyperonic and antiprotonic atoms) is included under **PHYSICS II, Nuclear Physics**. Also, high energy bombardment of targets where nuclear phenomena are the primary emphasis of the study is included under **PHYSICS II, Nuclear Physics**. If both nuclear and elementary particle phenomena are studied, both **Nuclear Physics** and **High Energy Physics** categories are assigned.)
645101	Electromagnetic Interactions and Properties
	(Photon interactions, photon production, and photon mediation (real or virtual) of any process and scattering by the electromagnetic force are always electromagnetic processes. All particles except neutrinos can take part in electromagnetic interactions. Particle decays mediated by the electromagnetic force (e.g., pions neutral) are included here. The lifetimes for electromagnetic decays are 10^{-16} to 10^{-20} seconds.)
645102	Weak Interactions and Properties
	(The neutrino always interacts weakly. All particles except photons can take part in weak interactions. Any particle decay yielding one or more leptons (electrons, neutrinos, muons minus, tauons minus, and their antiparticles) is a weak decay (includes most decays of elementary particles). The lifetimes for weak decays are 10^{-10} seconds or longer.)
645103	Strong Baryon-Induced Interactions and Baryon Properties
	(Includes strong particle decay of baryon resonances. The lifetimes for strong particle decays are 10^{-22} to 10^{-24} seconds.)
645104	Strong Meson-Induced Interactions and Meson Properties
	(Includes strong particle decay of meson resonances. The lifetimes for strong particle decays are 10^{-22} to 10^{-24} seconds.)
645200	*Particle Interactions and Properties—Theoretical*
	(Theoretical studies on the electromagnetic, weak, and strong interactions and on properties of particles and radiation quanta. Intermediate energy physics ($N\pi$ and Δ-1236 resonance dynamics in nuclear media, hypernuclei, and hyperonic and antiprotonic atoms) is included under **PHYSICS II, Nuclear Physics**.)
645201	General
645202	Electromagnetic Interactions and Properties
	(Photon interactions, photon production, and photon mediation (real or virtual) of any process and scattering by the electromagnetic force are always electromagnetic processes. All particles except neutrinos can take part in electromagnetic interactions. Particle decays mediated by the electromagnetic force (e.g., pions neutral) are included here. The lifetimes for electromagnetic decays are 10^{-16} to 10^{-20} seconds.)
645203	Weak Interactions and Properties
	(The neutrino always interacts weakly. All particles except photons can take part in weak interactions. Any particle decay yielding one or more leptons (electrons, neutrinos, muons minus, tauons minus, and their antiparticles) is a weak decay (includes most decays of elementary particles). The lifetimes for weak decays are 10^{-10} seconds or longer.)
645204	Strong Interactions and Properties
	(All particles except photons and leptons can take part in strong interactions. The lifetimes for strong particle decays are 10^{-22} to 10^{-24} seconds.)
645205	Strong Interactions and Properties, Baryon No. = 0
645206	Strong Interactions and Properties, Baryon No. = 1
645207	Strong Interactions and Properties, Baryon No. > 1

Category No.	Category Title
645300	*Particle Invariance Principles and Symmetries* (Space-time symmetries and internal symmetries of particles.)
645301	General
645302	Applications to Electromagnetic and Weak Interactions
645303	Applications to Strong Interactions
645400	*Field Theory* (Quantum field theory and quantum electrodynamics. Refer also to the **PHYSICS II, Theoretical Physics** category for related information.)
645500	*Scattering Theory* (General scattering theory only; the application of scattering theory to specific interactions or types of particles is included under sections according to type of interaction.)
650000	**PHYSICS II** (Physics is divided into two parts to enable proper subcategorization. Part II includes **Nuclear Physics; Radiation and Shielding Physics; Medical Physics; Solid State Physics; Theoretical Physics; Mathematical Physics;** and **Communication, Education, History, and Philosophy**.)
651000	**Nuclear Physics** (All scientific and technical aspects.)
651100	*Experimental Techniques* (Experimental techniques of general application in nuclear physics. Nuclear spectroscopic instrumentation is included under **INSTRUMENTATION**.)
651210	*Nuclear Properties and Reactions, A = 1-5, Theoretical*
651211	Mass, Abundance, and Binding Energy
651212	Radioactive Decay
651213	Energy Levels and Transitions
651214	Moments and Spin (Ground state only.)
651215	Nuclear Reactions and Scattering
651220	*Nuclear Properties and Reactions, A = 1-5, Experimental*
651221	Mass, Abundance, and Binding Energy
651222	Radioactive Decay
651223	Energy Levels and Transitions
651224	Moments and Spin (Ground state only.)
651225	Nuclear Reactions and Scattering
651310	*Nuclear Properties and Reactions, A = 6-19, Theoretical*
651311	Mass, Abundance, and Binding Energy
651312	Radioactive Decay
651313	Energy Levels and Transitions
651314	Moments and Spin (Ground state only.)
651315	Nuclear Reactions and Scattering
651320	*Nuclear Properties and Reactions, A = 6-19, Experimental*
651321	Mass, Abundance, and Binding Energy
651322	Radioactive Decay
651323	Energy Levels and Transitions
651324	Moments and Spin (Ground state only.)
651325	Nuclear Reactions and Scattering
651410	*Nuclear Properties and Reactions, A = 20-38, Theoretical*
651411	Mass, Abundance, and Binding Energy
651412	Radioactive Decay

Category No.	Category Title
651413	Energy Levels and Transitions
651414	Moments and Spin (Ground state only.)
651415	Nuclear Reactions and Scattering
651420	*Nuclear Properties and Reactions, A = 20–38, Experimental*
651421	Mass, Abundance, and Binding Energy
651422	Radioactive Decay
651423	Energy Levels and Transitions
651424	Moments and Spin (Ground state only.)
651425	Nuclear Reactions and Scattering
651510	*Nuclear Properties and Reactions, A = 39–58, Theoretical*
651511	Mass, Abundance, and Binding Energy
651512	Radioactive Decay
651513	Energy Levels and Transitions
651514	Moments and Spin (Ground state only.)
651515	Nuclear Reactions and Scattering
651520	*Nuclear Properties and Reactions, A = 39–58, Experimental*
651521	Mass, Abundance, and Binding Energy
651522	Radioactive Decay
651523	Energy Levels and Transitions
651524	Moments and Spin (Ground state only.)
651525	Nuclear Reactions and Scattering
651610	*Nuclear Properties and Reactions, A = 59–89, Theoretical*
651611	Mass, Abundance, and Binding Energy
651612	Radioactive Decay
651613	Energy Levels and Transitions
651614	Moments and Spin (Ground state only.)
651615	Nuclear Reactions and Scattering
651620	*Nuclear Properties and Reactions, A = 59–89, Experimental*
651621	Mass, Abundance, and Binding Energy
651622	Radioactive Decay
651623	Energy Levels and Transitions
651624	Moments and Spin (Ground state only.)
651625	Nuclear Reactions and Scattering
651710	*Nuclear Properties and Reactions, A = 90–149, Theoretical*
651711	Mass, Abundance, and Binding Energy
651712	Radioactive Decay
651713	Energy Levels and Transitions
651714	Moments and Spin (Ground state only.)
651715	Nuclear Reactions and Scattering
651716	Spontaneous and Induced Fission
651720	*Nuclear Properties and Reactions, A = 90–149, Experimental*
651721	Mass, Abundance, and Binding Energy
651722	Radioactive Decay

Category No.	Category Title
651723	Energy Levels and Transitions
651724	Moments and Spin (Ground state only.)
651725	Nuclear Reactions and Scattering
651726	Spontaneous and Induced Fission
651810	*Nuclear Properties and Reactions, A = 150–189, Theoretical*
651811	Mass, Abundance, and Binding Energy
651812	Radioactive Decay
651813	Energy Levels and Transitions
651814	Moments and Spin (Ground state only.)
651815	Nuclear Reactions and Scattering
651816	Spontaneous and Induced Fission
651820	*Nuclear Properties and Reactions, A = 150–189, Experimental*
651821	Mass, Abundance, and Binding Energy
651822	Radioactive Decay
651823	Energy Levels and Transitions
651824	Moments and Spin (Ground state only.)
651825	Nuclear Reactions and Scattering
651826	Spontaneous and Induced Fission
651910	*Nuclear Properties and Reactions, A = 190–219, Theoretical*
651911	Mass, Abundance, and Binding Energy
651912	Radioactive Decay
651913	Energy Levels and Transitions
651914	Moments and Spin (Ground state only.)
651915	Nuclear Reactions and Scattering
651916	Spontaneous and Induced Fission
651920	*Nuclear Properties and Reactions, A = 190–219, Experimental*
651921	Mass, Abundance, and Binding Energy
651922	Radioactive Decay
651923	Energy Levels and Transitions
651924	Moments and Spin (Ground state only.)
651925	Nuclear Reactions and Scattering
651926	Spontaneous and Induced Fission
652010	*Nuclear Properties and Reactions, A = 220 and Above, Theoretical*
652011	Mass, Abundance, and Binding Energy
652012	Radioactive Decay
652013	Energy Levels and Transitions
652014	Moments and Spin (Ground state only.)
652015	Nuclear Reactions and Scattering
652016	Spontaneous and Induced Fission
652020	*Nuclear Properties and Reactions, A = 220 and Above, Experimental*
652021	Mass, Abundance, and Binding Energy
652022	Radioactive Decay
652023	Energy Levels and Transitions
652024	Moments and Spin (Ground state only.)

Category No.	Category Title
652025	Nuclear Reactions and Scattering
652026	Spontaneous and Induced Fission
653000	*Nuclear Theory* (Studies not related to specific nuclides or groups of nuclides.)
653001	Nuclear Structure
653002	Nuclear Matter
653003	Nuclear Reactions and Scattering
653004	Nuclear Moments and Spin
653005	Radioactive Decay
653006	Spontaneous and Induced Fission
653007	Nuclear Models
654000	**Radiation and Shielding Physics**
654001	*Radiation Physics* (Basic physical characteristics of radiation or of radiation passage through materials not generally defined as shielding. This includes experimental or theoretical studies of range-energy relations, energy loss mechanisms, and absorption mechanisms. Neutron transport in macroscopic systems is included under *Neutron Interactions with Matter*. Radiation effects information is included under **BIOMEDICAL SCIENCES** and **MATERIALS**.)
654002	*Shielding Calculations and Experiments* (Neutron shielding and shielding material fabrication are included under *Neutron Interactions with Matter* and **MATERIALS**, respectively. Shield design is included under sections for the application, e.g., **PARTICLE ACCELERATORS** or **NUCLEAR REACTOR TECHNOLOGY**.)
654003	*Neutron Interactions with Matter* (Neutron absorption, scattering, thermalization, transport, etc., in macroscopic systems, including solution of the neutron transport equation and theoretical neutron transport in matter in general geometric configurations. Information on the reactions of neutrons with specific nuclei is included under the appropriate **Nuclear Physics** subcategory. Neutron kinetics in reactor lattices is included under **NUCLEAR POWER PLANTS** or **NUCLEAR REACTOR TECHNOLOGY**, depending on the emphasis.)
655000	**Medical Physics** (Interest is restricted to specific subject concepts indicated under the following subcategories.)
655001	*Radiation Protection Standards* (Standards written for a specific circumstance are included under appropriate sections.)
655002	*Radiation Source Calibration and Standardization* (Radiation source design and fabrication or radiation source preparation are included under **ISOTOPE AND RADIATION SOURCE TECHNOLOGY** or **CHEMISTRY**.)
655003	*Dosimetry* (Calculations of dose distributions in tissue-equivalent materials or phantoms. Dose determinations in biological materials are included under **BIOMEDICAL SCIENCES, APPLIED STUDIES**.)
656000	**Solid-State Physics** (Theoretical treatments only; solid-state properties of specific materials are included under **MATERIALS**. Interest is restricted to DOE-supported research.)
656001	*Solid-State Plasma* (All experimental and theoretical aspects; thermonuclear plasmas are included under **FUSION ENERGY**. Interest is restricted to DOE-supported research.)
656100	**Superconductivity** (Basic theories and phenomenological studies only; properties of specific superconducting materials are included under **MATERIALS**, superconducting equipment and devices are included under **ENGINEERING** or categories for the application, e.g., **PARTICLE ACCELERATORS, ELECTRIC POWER ENGINEERING, DIRECT ENERGY CONVERSION**, etc.)
656101	*General Theory*
656102	*Acoustic, Electric, Magnetic, Optical, and Thermal Phenomena*
657000	**Theoretical Physics**

Category No.	Category Title
657001	*General and Miscellaneous* (Interest is restricted to DOE-supported research.)
657002	*Classical and Quantum Mechanics*
657003	*Relativity and Gravitation* (Basic studies of gravitation theory and general relativity theory. Gravitational collapse, gravitational wave detection, and tests of general relativity are included under **PHYSICS I, Astrophysics and Cosmology.**)
657004	*Acoustics* (Interest is restricted to DOE-supported research.)
657005	*Optics* (Interest is restricted to DOE-supported research.)
657006	*Statistical Physics and Thermodynamics* (Interest is restricted to DOE-supported research.)
657007	*Electricity and Magnetism* (Interest is restricted to DOE-supported research.)
658000	**Mathematical Physics** (Restricted to mathematical treatment of concepts involved in the other physics categories; should usually be cross-categorized to one or more of those categories.)
659000	**Communication, Education, History, and Philosophy** (Interest is restricted to DOE-supported research.)
700000	**FUSION ENERGY** (Fusion information of a general nature or covering several aspects, e.g., bibliographies, books, DOE or other programmatic efforts, etc.)
700100	**Plasma Research**
700101	*Confinement and Heating* (Studies on plasma lifetime, energy loss, and energy balance in fusion devices; also plasma physics of heating studies such as beam injection and high-frequency techniques. Related engineering such as neutral-beam sources, microwave equipment, or pellet injectors is included under **Fusion Power Plant Technology,** *Fuel, Heating, and Injection Systems.*)
700102	*Diagnostics* (Diagnostic techniques and apparatus.)
700103	*Kinetics—General* (General plasma physics studies when it is not possible to assign a more specific category and for papers that may be considered both experimental and theoretical.)
700104	*Kinetics—Experimental* (Only experimental studies on plasma physics.)
700105	*Kinetics—Theoretical* (Only theoretical studies on plasma physics.)
700106	*Production* (Includes such studies as preionization techniques for start-up and laser-produced plasmas in magnetic geometries. Laser-produced plasmas for inertial confinement are included in **Fusion Power Plant Technology,** *Inertial Confinement Technology.*)
700107	*Instabilities* (Any aspect of plasma microinstabilities and macroinstabilities in fusion devices.)
700108	*Wave Phenomena* (All types of plasma wave and electromagnetic wave interaction and propagation studies.)
700200	**Fusion Power Plant Technology** (Planning and conceptual design studies, i.e., pure fusion and hybrid reactors.)
700201	*Blanket Engineering* (Physics and engineering related to blankets, first wall, shielding, and impurity control.)
700202	*Magnet Coils and Fields* (Experiments, design analyses, and design code coordinating activities related to magnets and magnetic-field configurations, e.g., divertors.)
700203	*Power Supplies, Energy Storage* (Design and performance analyses for any power supply or energy storage system associated with a fusion device.)
700204	*Cooling Systems* (Studies on heat transfer or system components.)

Category No.	Category Title
700205	*Fuel, Heating, and Injection Systems* (Studies on tritium, deuterium, and boron-11 for use as fuel, including processing, inventories, and availability; also plasma guns, microwave equipment, neutral-beam sources, etc., for use in beam injection.)
700206	*Environmental Aspects* (General studies on environmental impacts or safety of pure fusion or hybrid reactors; safety aspects in the fuel handling and processing cycle; and management, processing, disposal, or storage of radioactive wastes. See also **NUCLEAR FUELS, Waste Management.** Hazards or safety in tritium containment and recovery; for design, development, etc., of shipping containers, see **ENGINEERING.**)
700207	*Power Conversion Systems* (Studies on MHD topping cycles, direct energy converters, gas turbines, etc.)
700208	*Inertial Confinement Technology* (All studies related to laser fusion, electron-beam fusion, ion-beam fusion, and impact fusion. Any aspect of the driver system, target, or beam transport and modeling is included.)
700209	*Component Development and Materials Testing* (Studies of reactor components such as vacuum systems, limiters, etc.; any materials studies related to fusion research; production of synthetic chemical fuels.)
990000	**GENERAL AND MISCELLANEOUS** (Interest restricted to DOE-sponsored research or energy-related information not more appropriately categorized elsewhere. Therefore, these should not be used as secondary categories.)
990100	**Management**
990200	**Mathematics and Computers** (Includes DOE-supported research in general. Specifically includes all aspects of supercomputers—the large, fast, state-of-the-art general-purpose computers—and associated computer language theory and development.)
990300	**Information Handling**
990400	**Law**
990500	**Civilian Defense**

INDEX

This index is included to assist the user in quickly locating keywords or phrases in the category titles or category scope notes. It *should not* be used when assigning categories or when searching the Energy Data Base; the text should always be consulted before performing either function.

SUBJECT INDEX

Subject terms used in this index are, with a few exceptions, those used by the U. S. Department of Energy in the publication **ENERGY DATA BASE SUBJECT CATEGORIES AND SCOPE** (Appendix III). If the desired subject term is not in the INDEX, check Appendix III for additional subject possibilities or try related subject areas.

The numbers following the subject headings refer to entry numbers in the "Master List" section. A * following an entry denotes that the periodical generally incorporates a policy of peer review. It should be noted that many publications which do not practice peer review are promoting and publishing research which is so innovative or new that it cannot be subjected to a normal review process. Consequently, users are encouraged to view entries other than those with an *.

Multiple subjects are frequently strung together. This has been done to avoid redundancy of entry numbers when two or more subjects having identical entry numbers appear sequentially. While the reason behind this is to save the user time, it is important to read the subject categories carefully.

About the Compiler

ROBERTA A. SCULL is Director of Information Services for the Center of Energy Studies at Louisiana State University, Baton Rouge, and a former member of the Depository Library Council to the Public Printer of the United States (1979-1982). She has compiled *Bibliography of United States Government Bibliographies, 1968-1973*, and *Bibliography of United States Government Bibliographies, 1974-1976*, and has contributed to *The Basic Business Library: Core Resources*, *Reference Services Review*, *Special Libraries*, and other library publications.

www.ingramcontent.com/pod-product-compliance
Lightning Source LLC
LaVergne TN
LVHW011904200926
848484LV00002B/530